Your IntroPsych Journey Begins Here

MW01518687

With over
3000
students registered

IntroPsych is McMaster's most
POPULAR
course

100+
Tutorials

Creating an
intimate
classroom environment

TA's Average
4.6 / 5
on evaluations

Delivering
top-ranked
performances according to students

225
office hours

Guaranteeing you
Resourceful
assistance when you need it

Award-Winning*
Instruction

Ensuring the highest
Quality
of education

Blended Learning

Providing students with an
Innovative
learning experience

320+
?
practice questions

Helping you get
Prepared
for the final exam

26,560
Graduates

Gearing up for
Successful
futures in psychology and beyond

A Better, Smarter, YOU

Ready to tackle the world

*2010 President's Award Excellence in Course and Resource Design
*2010 Vice-President Research Innovator of the Year
*2012 Residence Life Campus Partner of the Year
*2014 nominated for 3M Canada National Teaching Fellowship

Created By Nabil Khaja

INTROPSYCH 1X03/1N03/1F03

Official Course Handbook Fall 2014

Web Module and Tutorial Guide

INTROPSYCH.NET

Your one-stop course companion website.
Includes:

- Practice exam questions
- Expanded glossary of terms
- Academic advice
- Departmental information
- Student life directory
- And more!

NOTICE:
These learning materials are to be used as a guide to assist you with understanding the module material – they are not a replacement to viewing the online modules, attending your tutorials, or taking your own notes. Your grade will be determined by your full participation in all aspects of this course.

If you find any errors or omissions within this document, please feel free to email us (intropsych@mcmaster.ca) and we will be sure to have the issue rectified for the next release of this handbook.

You may not reproduce or distribute any portion of this document

This handbook was made possible by the tireless efforts of our handbook team, consisting of:
Dr. Joe Kim: Course Instructor
Dr. Michelle Cadieux: Course Coordinator
Andrew LoGiudice: Special Contributor
Nabil Khaja: Special Contributor
Chris McAllister: Media Specialist
Greg Atkinson: Media Specialist

© Department of Psychology, Neuroscience & Behaviour
1280 Main Street West • Psychology Building
Hamilton, Ontario L8S 4K1
Phone 905.525.9140 x 24428
Email: intropsych@mcmaster.ca
Web: http://www.intropsych.net

Contents

Quick Start Guide

Before your first tutorial

1. Read the **ENTIRE** course outline (Pages 6-22).
2. Write the important dates in your calendar (See page 22 for dates).
3. Make a checklist of the course features (pages 18-19) and check out each one!
4. Complete the practise Quiz on Avenue to help you get comfortable with the format of online testing.

During the semester

1. Watch the weekly web modules **BEFORE** your tutorial. Take _effective_ notes using the outline and slides provided in this handbook as a guide.
2. Use a separate notebook for additional notes and write down any questions you have or concepts that were not clear and bring them to tutorial to ask your TA.
3. For weekly web modules, look up unfamiliar words in the glossary and review to get a solid foundation. As with most university courses, the psychology knowledge base you are expected to understand is cumulative and small gaps can turn into large gaps.
4. Complete the assigned readings (textbook/journal articles) to gain context, background, and additional examples to consolidate your knowledge.
5. Study Groups - Join one or start one with tutorial members or peers on AVE.
6. Complete the weekly activities and questions in the handbook (you may consider using a separate piece of paper to make sure you have enough space). These can be a great starting point for your study group.
7. After completing each week's Pre-Quiz and Quiz, be sure to review the feedback on your results to see where you when wrong and right.

McMaster Student Absence Form (MSAF)

What is it?
This is a self-reporting tool that allows you to report absences for up to 5 days due to illness. You are limited to 1 per semester so use it wisely. The MSAF form should be filled out _immediately_ upon your return to class and it is _your_ responsibility to follow up with your instructor to discuss the nature of the accommodations. If you miss more than 5 days or exceed 1 request per semester, you must contact your Associate Dean's Office (Faculty Office). At this point, you may be required to provide additional supporting documentation.

Please use intropsych@mcmaster.ca as the contact email on your MSAF

NOTE: it is _your_ responsibility to catch up on missed work! It is a good idea to get peer contacts during the first week of classes so you can share notes and catch each other up on missed work.

Where do I go to fill this out?
Go to the McMaster website and either search MSAF or type in http://www.mcmaster.ca/msaf/ in the search bar. At this point, read the page and click agree. From there you will be guided through the process in filling out the form.

Absences/Missed Quiz/Tutorial/Exam
If you must miss a quiz, tutorial, or exam due to illness or other situations, the Course Coordinator requires the permission of your own Associate Dean's office to excuse your absence. Contact them first about your situation. You may require documentation (e.g. doctor's note, wedding invitation). The Associate Dean's office will make a decision and forward the dates of any excused absence to the Course Coordinator without details of your illness/situation. Therefore, **go to the Associate Dean's office of your faculty FIRST!**

How to Use Google Scholar

Google scholar is a great tool to find peer reviewed journal articles. It is quick and simple. Below, we have outlined a few tips to make it even simpler.

How do I get there?

Easy! Type in http://scholar.google.ca, or just type "scholar" into Google.

I can't seem to access any articles, what do I do?

Most journals require you to purchase an article before you can access it. Luckily, McMaster has purchased these articles for us! All you have to do is click on "scholar preferences" at the top right of the window. Where it says "Library Links" enter "McMaster" then scroll to the bottom and save preferences. Simple!

I am trying to find articles about a general topic, what do I do?

Use the search box to search articles by their journal, author(s), title, or topic. Another search option is to click on "advanced scholar search" beside the search box. This will allow you to search for specific journals, authors, and titles. You can even search for articles about a particular topic published within a particular time frame and published in a particular journal. Play around with it and get used to this tool. It will come in handy not just this semester, but for much of your university career.

How to Use this Handbook

What Is Your Handbook?

Your handbook is the official courseware compiled by the Introductory Psychology instructional team and is designed to facilitate your academic experience in Psych 1X03/1N03. This handbook is continually revised to ensure that students have effective study aids and up-to-date information for the course. Your handbook is not a textbook. It does not contain mandatory readings that are required for you to pass the course. However, it does have a wealth of resources to help you reach your academic goals in Psych 1X03/1N03.

How To Use Your Handbook Effectively

Taking Notes: The inclusion of important graphics is meant to facilitate your learning so that you do not have to recopy all of the information from the web lectures. In addition, ample space has been provided beside each key slide for you to take notes in your own words. Furthermore, the skeleton notes are ordered sequentially, thus providing some additional structure to your note taking strategies. Copying down the script from the web lectures word for word is not an effective way of studying. Instead, watch the web modules once, reflect on the material you've just viewed, and then go back to re-watch the lectures while taking notes on the concepts you're struggling with.

Activities and Practice Questions: Doing activity and practice questions at the end of each module and checking your answers against the answer key is an effective way to learn and assess your understanding of important concepts. We encourage you to review the answers after finishing the questions to see if you're on the right track. Even if you have selected the correct answer, make sure to read through the explanations provided to ensure that your logic is accurate and to have trickier concepts explained again using different words.

Concept Maps: The concept maps are designed to help students see the "big picture". You can use concept maps as a quick pre-quiz review or to check if you have accurately understood all of the connections between the different elements in the module. As you begin studying for your final exam, you may also wish to use your concept map to check if you have forgotten a key concept in the course.

Sample Quiz Question Explained: Make sure you answer the question yourself and write this answer down before reading the explanation provided. This way you can accurately assess whether or not you know the material before doing your pre-quiz or Avenue quiz. Additionally due to the feedback provided, if you've selected an incorrect answer, you may be able to identify why you did not choose the correct answer and how to avoid making the same mistake on future quizzes. By critically evaluating your approach and performance on multiple-choice questions, you will find it easier to reach your desired level of performance on our quizzes and in the course. This may also be very helpful for your tests in other courses.

Glossary: Use your glossary as a quick reference for the key terms without having to re-watch modules to obtain definitions. Your glossary clearly defines the key terms and even provides you with new examples. You can test your knowledge of these terms by covering the definitions, writing down your answers to a specific term, and then checking to see if you are correct. Conversely, you could cover the terms, read the definition and see if you can accurately match the definition with its corresponding term.

Campus Resources

The following are some of the resources available to McMaster University students. Read over each description and familiarize yourself with what is available!

Student Accessibility Services

Student Accessibility Services offers various supports for students with disabilities. They work with full-time, part-time, and prospective students. SAS assists with academic and disability-related needs, including learning strategies, note-taking, assistive technologies, test & exam administration, accommodations for courses, groups, and events.
Website: http://sas.mcmaster.ca/
Phone: 905-525-9140 ext. 28652
Teletypewriter (TTY): 905-528-4307
Email: sas@mcmaster.ca
Office: McMaster University Student Centre (MUSC) - Basement, Room B107

Student Wellness Centre

The Student Wellness Centre offers medical & health services, personal counselling, and mental health services.

Medical & Health: The Student Wellness Centre provides a wide range of health services for students and will act as their personal health care provider throughout their studies at McMaster. Appointment bookings should be made ahead of time.
Personal Counselling & Mental Health: At some point, almost everyone experiences major concerns that may interfere with their success, happiness, and satisfaction at university. Common concerns are relationships, mood disorders, learning disabilities, body image, anxiety, and depression. The Student Wellness Centre provides experienced counsellors for bookings and emergency appointments.

Website: http://wellness.mcmaster.ca/
Phone: 905-525-9140 ext. 27700
Email: wellness@mcmaster.ca
Office: MUSC B101

McMaster Institute for Innovation and Excellence in Teaching and Learning (MIIETL)
This institute is designed to enable success in teaching and learning. Their activities include both general and discipline-specific approaches to the promotion of successful learning in all its forms and contexts.

Website: miietl.mcmaster.ca
Phone: 905-525-9140 ext. 24540
Office: Mills Library L504

Academic Advising in PNB

Ann Hollingshead is the academic advisor for anyone in the PNB department at McMaster. She has a lot of knowledge about upper-year courses and can help you make the best decisions about what courses to take. Ann is available Monday – Friday 9:00am-5:00pm for drop-in help or by appointment.

Email: hollings@mcmaster.ca
Phone: 905-525-9140 ext. 23005
Office: PC 109

Note: For academic advising for other department, visit your faculty's website (Social Science, Science, Engineering, etc.) for more information.

Psychology 1X03/1N03 Course Outline - Fall 2014

Course Staff	Location	Office Hours
Dr. Joe Kim Instructor	PC/106	Posted weekly on Avenue
Dr. Michelle Cadieux Course Coordinator	IntroPsych Office PC/110	Posted weekly on Avenue
Rachael Barnette & **Ahmed Labib** Head TAs	PC Lobby	Posted weekly on Avenue

All correspondence regarding this course should be sent to: ***intropsych@mcmaster.ca*** using your McMaster email and **NOT** your Avenue account. If you have additional questions regarding course material you have several options:

- Ask Course Staff or TAs during Office and Lobby hours, posted weekly on Avenue
- Ask your personal TA during tutorial
- Join the active discussions on Avenue forums.

You may also call the IntroPsych office at X24428 during office hours. Please note that **phone messages will not be returned.** If you have a request, please see in person during office hours or send an email to intropsych@mcmaster.ca. Please note that during busy periods, it may take up to 48 hours to return your email message. Please be patient!

In addition to the course staff, you have been assigned to a tutorial section with your personal **Teaching Assistant (TA)** who will lead your group through weekly discussions, activities and questions.

Course Description

This course introduces a scientific framework to explore important questions in psychology, neuroscience and behaviour. Using psychological research methods to understand learning, cognition, and social psychology, this course helps learners to develop skills to integrate, evaluate and examine information that is useful in applied settings. The intended learning outcomes are to:

- Integrate knowledge from research design, learning, cognition, personality and social psychology
- Apply the language of psychology in real-world settings and connect with current events
- Critically evaluate scientific information, data and research methodologies
- Discover how psychological theories help assess, predict or change human behaviour

Evaluation

Your final grade in Psychology 1X03 will be determined by the following measures:

Tutorial Participation	10%
Avenue Quizzes	30%
Final Examination	60%
Research participation (optional)	5%

Tutorial Participation (10%)

Weekly tutorials are an important part of the course and contribute to 10% of your final grade. Your TA will expect **active** participation to create a dynamic learning environment. If you have specific issues with this process you must speak with your TA as soon as possible. **Every three weeks**, your TA will assign you a tutorial grade out of 10 using the rubric below as a guide.

EVALUATING CONTRIBUTION

ATTENDANCE	Excellent	Satisfactory	Poor
3 of 3 classes	8-10	6-7	4
2 of 3 classes	6-8	2-4	0-2
1 of 3 classes	4	0	0

Excellent	Regularly makes thoughtful contributions
Satisfactory	Occasional makes valuable contributions
Poor	Little to no participation

Note: students who regularly attend tutorials but make little or no contribution to discussions cannot receive a grade higher than 4 out of 10. Therefore, it is essential that you **actively participate** if you wish to earn a high participation grade. Your TA can help you with suggestions for demonstrating active participation

Tutorial i<clicker Bonus Points

On some weeks, your TA will give you the opportunity to earn i<clicker Bonus Points during a tutorial. These will take the form of a short quiz at the beginning of class. Depending on the quiz, you will be allowed to work in groups or individually. However, you will always be asked to submit an individual answer. All tutorial bonus quizzes require an i<clicker to complete. They can be purchased at the bookstore. You **must** bring your i<clicker to every tutorial if you wish to participate in these bonus opportunities. Each correct answer will earn you bonus points. At the end of the term, the total number of points earned will lead you to the following rewards:

Points	Reward
22 and above	Drop lowest tutorial participation grade
30 and above	Drop lowest tutorial participation grade + drop lowest quiz grade + enter draw for lunch with Dr. Kim (seriously)

Register your i<clicker

To register your i<clicker please go to http://www1.iclicker.com/register-your-remote/register-clicker/

Your Student ID is your MacID, which is typically the first part of your McMaster email address. Failure to bring your personal i<clicker to tutorial or failure to register your i<clicker properly will result in a zero being assigned to the relevant Tutorial Bonus Point quiz.

Avenue Quizzes (30%)

There are 11 weekly online Avenue Quizzes during the semester which will cover material from the assigned Web Modules AND assigned readings from the Textbook or Journal articles. For example, Quiz 1 will contain material covered from the Introduction Web Module and the Textbook Chapter 1 Reading. Each Quiz is "open book" and you may collaborate with your peers but you may NOT post questions. Avenue Quizzes are an opportunity to assess and consolidate your knowledge of the week's content in preparation for the Final Exam where you will be working independently and without access to supporting resources.

Each Quiz will cover the web module and textbook readings from the same week as covered in tutorial. Each Avenue Quiz will consist of 10 multiple-choice questions. Avenue Quizzes will be made available online every Friday at 6AM and will promptly close on Saturday at 6AM. You will have 20 minutes to complete each quiz once you open it. After closing on Saturday, the Quiz will be reviewed and grades will be released on the following Tuesday.

Quiz questions are designed to go beyond mere recall and challenge you to apply and demonstrate your comprehension. In other words, simply memorizing terms will not lead to a favourable grade. To help you prepare and assess your study, you will have a **Pre-Quiz** each week (released on Monday) that will be graded immediately and will provide you with feedback on why your chosen option was correct/incorrect. You can review the completed Pre-Quiz under the Quizzes tab on Avenue. The Pre-Quiz serves as an excellent learning tool as it is drawn from the same question bank as your actual Friday Quiz, but does not officially count for grades. Note that the Pre-Quiz **must** be completed to gain access to the weekly Avenue Quiz.

Internet Problems

Internet issues can happen. We always recommend that you complete your quizzes on campus where a reliable Internet connection is assured. While we do not accommodate for individual Internet issues, we can grade your quiz manually if you take screen shots. Please ensure that all photos have your name and timer in the shot.

Final Exam (60%)

A cumulative Final Exam will be written in December as scheduled by the Registrar's Office. If you choose to complete the optional research participation option (see below), the weight of your final examination will be reduced from 60% to 55%. The Final Exam covers material presented in web modules, tutorials, and assigned readings from the entire term.

IntroPsych Discovery Challenge

The IntroPsych Discovery Challenge is an extracurricular activity that pits tutorials against each other in a course-wide competition to earn additional bonus points. Each week, IntroPsych students will be presented with a challenging word puzzle to solve that extends beyond the content covered in class. Using the knowledge they have acquired during the previous week, as well as library and internet resources, tutorials will venture to submit a correct answer to that week's puzzle. A breakdown of the schedule for each week is as follows:

Monday through Thursday: Web modules and tutorials on the week's topic

Friday: Online Avenue Quiz for the week's topic

Sunday 9:00PM: Discovery Challenge puzzle is released on the topic tested the previous week. The puzzle will be posted on the Course Announcements section of Avenue to Learn. The puzzle will require students to go into more detail on the topic than was covered in the web module, and students will likely need to use online and library resources to learn and apply their knowledge to solve the puzzle. The puzzle alone should not provide enough information to answer, only enough to support an educated guess to the answer provided additional information was researched. We expect, and will encourage, students to repost the puzzle on their private tutorial-specific discussion boards to discuss possible answers.

Monday 9:00PM: Clue #1 will be posted on Course Announcements. This clue should enable students to answer the puzzle, provided additional research was completed.

Tuesday 9:00PM: Clue #2 will be posted on Course Announcements. The two clues together should make the puzzle much easier to answer.

Wednesday 9:00PM: Any final answers must be submitted by 9:00 PM.

Tutorials have 3 time windows to submit answers:
1) After the puzzle is released but prior to release of Clue #1
2) After Clue #1 but prior to the release of Clue #2
3) After Clue #2 but prior to Wednesday at 9 PM.

Only **ONE** answer can be submitted per time window per tutorial and only the **FIRST** answer submitted by a tutorial within any of these time windows will be considered. There is no penalty for incorrect answers

To submit answers, each tutorial will be given a unique ID and password used to login to IntroPsych.net. Students within a tutorial are expected to discuss possible answers to the puzzle and, once they are confident, **elect somebody to represent their tutorial** in submitting their answer on IntroPsych.net. The student will login using their tutorial ID and password, and submit their student number alongside their answer to the puzzle. All tutorials submitting an answer by 9:00 PM will receive points, with more points earned if correct and submitted before the release of clues.

Answers will be scored according to this breakdown:

Correct answer before Clue #1	5 points
Correct answer before Clue #2	4 points
Correct answer before Wed 9 PM deadline	3 points
Any answer submitted	1 point

13

The first 5 tutorials to submit a correct answer will score additional points:

1st to answer correctly	+2 points
2nd to 5th to answer correctly	+1 points

For example: If you are the first tutorial to correctly answer before clue 1, your tutorial gets 7 points (5 + 2). But if you are the 6th or later, your tutorial will only get 5 points. If no tutorial has correctly answered it before clue #1 and you are the second tutorial to correctly answer it after clue #1 is released, you will get 5 points (4 + 1). However, tutorials 6th or later will only receive 4 points.

Thursday 9:00PM: The answer to the week's puzzle will be posted, along with an explanation of the answer and clues. At this time, the tutorials scoring in the top 5 for the week will be announced on Avenue to Learn. A leaderboard of all tutorials' total points will also be visible online.

At the end of the semester, each student in the tutorial group that finishes with the most points in the course **will earn a letter grade-bump for their final mark in the course!** For example, all A's will become A+'s and all C+'s will become B-'s. **However, this conversion will not convert any failing grades to passing grades.**

At the end of the term, all students in any tutorial holding more than 10 points will have their lowest AVENUE Quiz dropped from their grade. All students in any tutorial holding more than 25 points will have their lowest two AVENUE Quizzes dropped from their grade. (Note: This means that students will be able to drop one AVENUE Quiz simply by participating each week —even if they do not get any of them correct.)

Discovery Challenge is an extracurricular activity not meant to be discussed during tutorial, and should only involve IntroPsych students. **TA's are not allowed to contribute to any in-person or online discussion of the puzzles.** If students have questions or concerns about the format or answer submission system, they can post on the Discovery Challenge discussion board.

Research Participation Option

You have the option to reduce the weight of your Final Exam from 60% to 55% by completing and attaining **two credits** of research participation with the Department of Psychology, Neuroscience, and Behaviour. In addition to providing you with extra credit, the research participation option allows you to take part in some exciting research taking place right here at McMaster and observe how psychologists conduct their studies.

The system that the department uses to track research participation is called Sona, which can be accessed at **mcmaster.sona-systems.com**. To access Sona for the first time, select the "New Participant?" option at the bottom-left of your screen and enter your name, student number, and McMaster email address (for security reasons, *only* your McMaster email address may be used). After a short delay, you will receive an email from Sona with a username and temporary password that you can use to access the website. You should change your temporary password to something more memorable by selecting "My Profile". Make sure your student number is entered correctly! **Note: You must activate you McMaster ID before you can create a Sona account. To activate your ID, please go to www.mcmaster.ca/uts/macid**

Completing Your Research Participation Credit

When you log into Sona for the first time, you will be prompted to chose a course. Please select Psych 1X03 from the list. You will also be asked to fill out a short survey. This information is used filter out any experiments for which you are not eligible.

To register for an experiment, select "Study Sign-Up" from the main Sona page. You will be presented with a list of currently available experiments, with a short description given about each. Before selecting an experiment, be sure to read the description carefully, making special note of any specific criteria for participation (for example, some experiments only allow females to participate, while others may require subjects who speak a second language). When you have found an experiment that you would like to participate in, select "View Time Slots for this Study" to view available timeslots, then select "Sign-Up" to register for a time that fits your schedule. You will receive a confirmation email with the details of your selection. Be sure to write down the experiment number, experimenter name, location, and telephone extension from this email.

After you have completed an experiment, you will be given a purple slip verifying your participation. This slip is for your records only—in the event that an experiment is not credited to your Sona account, this slip is your proof of participation. Shortly after completing an experiment, you should notice that your Sona account has been credited by the experimenter.

Additional Notes
- You must complete two full hours of experiments, and no less, if you wish to earn the 5% credit.
- If you do not wish to participate as a research subject for any reason, you may still earn your research participation credit by *observing* two hours of experiments. If you would like to choose this option, please see the course coordinator, Dr. Michelle Cadieux, in PC 110.
- If you fail to show up for two experiments, you will lose your option to complete the research participation credit. If you know in advance that you will be unable to attend a scheduled experiment, please contact the experimenter.

Privacy and Conduct

In this course we will be using AVENUE for the online portions of your course. Students should be aware that when they access the electronic components of this course, private information such as first and last names, user names for the McMaster e-mail accounts, and program affiliation may become apparent to all other students in the same course. The available information is dependent on the technology used. Continuation in this course will be deemed as consent to this disclosure. If you have any questions or concerns about such disclosure, please discuss this with the Course Coordinator.

All posts on discussion forums should be polite and refrain from derogatory and unacceptable language.

A Note about Academic Honesty

Academic dishonesty consists of misrepresentation by deception or by other fraudulent means and can result in serious consequences, e.g. the grade of zero on an assignment, loss of credit with a notation on the transcript (notation reads: Grade of F assigned for academic dishonesty), and/or suspension or expulsion from the university. It is the student's responsibility to understand what constitutes academic dishonesty. For information on the various kinds of academic dishonesty please refer to the Academic Integrity Policy, specifically Appendix 3 at: **http://www.mcmaster.ca/univsec/policy/AcademicIntegrity.pdf**

The following illustrates only three forms of academic dishonesty:
- Plagiarism, e.g. the submission of work that is not one's own or for which other credit has been sought or obtained;
- Improper collaboration; or,
- Copying or using unauthorized aids in tests or examinations.

Changes during the term

The instructor and university reserve the right to modify elements of the course during the term. The university may change the dates and deadlines for any or all courses in extreme circumstances. If either type of modification becomes necessary, reasonable notice and communication with the students will be given with explanation and the opportunity to comment on changes. It is the responsibility of the student to check their McMaster email and course websites weekly during the term and to note any changes.

A Note About Note Taking

Students often wonder (and worry) about how extensive their notes should be. This handbook provides outlines with key points and slides reproduced from the web modules to guide your own note taking. There really is no substitute for doing this yourself to learn the material. If, however, you can refer to your notes and answer the practice questions that follow the handbook outlines, you should find yourself in good shape for the weekly quizzes and the exam to come.

A Note About Tests

With practice questions in this handbook, Tutorial Bonus Point questions, Pre-Quiz and Quizzes, you might be wondering "why are there so many tests?!". The simple answer is that testing has been shown to be the most effective way to learn information in the long term.

Retrieval-Enhanced Learning

Many students likely view testing as a negative necessity of their courses and would prefer to have as few tests as possible. Thinking about testing this way is due to years of experiencing tests as a high-stakes assessment of learning. This is unfortunate given the fact that testing improves learning. Over the past hundred years research on the characteristics of human learning and memory has demonstrated that practice testing enhances learning and retention of information (e.g., Dunlosky et al., 2013; Roediger & Karpicke, 2006a). Practice testing can take many forms. It can refer to practicing your recall of information by using of flashcards, completing practice problems or questions in a textbook, or completing low-stakes tests as part of a course requirement. This principle was the primary motivation for redesigning the IntroPsych course to have weekly, low stakes quizzes.

An excellent example of the power of testing memory comes from a study by Roediger and Karpicke (2006b), wherein undergraduate students were presented with short, educationally relevant texts for initial study. Following initial study students either studied the material again, or took a practice test. A final test was taken after a short retention interval (5 minutes) or long retention interval (2 days). After a short retention interval restudying produced better recall than testing (81% vs. 75%). However, with the long retention interval testing produced significantly better recall than restudying (68% vs. 54%). Thus, after two days performance declined 27% for students who restudied the material, but only 7% for students that practiced recall.

Interestingly, providing students with the correct answer feedback after a test enhances the positive effect of testing. With feedback, learners are able to correct errors, and maintain their correct responses. Moreover, taking a test and reviewing feedback can enhance future study sessions. Research shows that when a student takes a test before restudying material, they learn **more** from the restudying session than if they restudy without taking a test beforehand (e.g., Karpicke, 2009). This is called test-potentiated learning.

Why does testing improve retention of information and how can I use it?

Explanations for the positive effects of testing focus on how the act of retrieval affects memory. Specifically, it is suggested that retrieving information leads to an elaboration of memory traces and the creation of additional retrieval paths. Together these changes to memory systems make it more likely that the information will be successfully retrieved again in the future. This suggests that testing is not just an assessment tool, but also an effective learning tool.

As a student in this course you can take advantage of retrieval enhanced learning each week in preparation for your weekly Quiz. This begins with studying web module content early in the week (e.g., Sunday or Monday). You can then engage in retrieval practice as a form of review after your initial study session. At this point you should be ready to complete the Pre-quiz and review the feedback (by Wednesday or Thursday). This gives you the opportunity to take advantage of test-potentiated learning when you review content again before completing your Quiz on Friday. This suggested schedule of studying, and incorporation of retrieval practice will help you learn and retain the course content. Engaging in this process each week enhances your long-term memory for course content and therefore advances your preparations for the final exam!

Suggested further reading:

1. Dunlosky, et al., (2009). *Psychological Science in the Public Interest, 14*(1), 4-58.
2. Karpicke, J. D. (2009). *Journal of Experimental Psychology: General, 138*, 469–486.
3. Roediger, H. L., & Karpicke, J. D. (2006a). *Psychological Science, 17*, 249–255.
4. Roediger, H. L., & Karpicke, J. D. (2006b). *Perspectives on Psychological Science, 1*, 181–210.

Welcome to Psychology 1X03

Welcome to PSYCH 1X03: Introduction to Psychology, Neuroscience & Behaviour, one of two IntroPsych courses offered at McMaster University (PSYCH 1XX3 is offered in Term 2). Your IntroPsych course follows in the tradition of McMaster University's long-standing reputation of excellence in innovative teaching and learning. In this course, you will experience a unique blended learning model that combines online learning technology with traditional face-to-face instruction. On your way to the weekly Quizzes and Final Exam, there are several resources available to help you master the curriculum:

Course Handbook: Your course handbook contains valuable information regarding course structure, outlines, and guides for web modules and tutorials.

Course Textbook: Your course textbook can be purchased at the McMaster Bookstore and it contains essential readings with testable material for the course.

IntroPsych.net: There are many supplementary resources that have been specially developed to compliment the handbook at IntroPsych.net including practice tests, study aids, interactive glossary, information about course events, university's services, tips for academic success, and student life information. A portion of the proceeds from this courseware go toward the development and maintenance of IntroPsych.net

Avenue to Learn: Your primary course content will be delivered through the AVENUE learning management system located at **http://Avenue.mcmaster.ca**. AVENUE allows you access to weekly web modules, course announcements, discussion forums, and grade records. To access AVENUE, use your MacID and password. Below are some of the features of AVENUE.

Web Modules: The most unique feature of IntroPsych at McMaster is the way you receive your primary course content—it's all online! You can access the web modules from the library, your room, or anywhere you have an internet connection. The interactive web modules feature audio, video animations, and vivid graphics. Check out the many advanced features that allow you to interact with the content according to your personal learning style. Use the navigation tools and integrated search function to move about the module. You can test your knowledge with checkpoints, learn more about faculty related research through *Beyond IntroPsych*, leave your comments with the *Shout Wall*, take a poll, and interact with fellow students and course staff on *Live Chat*.

New web modules are released every Monday at 6PM for the *following* week's tutorials. Once a web module is released, it stays up all year for you to reference. However, be sure to view the assigned web modules **before** you arrive at your weekly tutorial session to stay on schedule and actively participate.

Discussion Boards: More extended topic discussions are available on the AVENUE Discussion Board. Join an existing discussion or start a new thread. Our discussion boards are consistently the most active compared to any other course on campus, so jump right in with your opinion!

Discover Psychology: Science You Can Use: If you are interested in pursuing a program in psychology or are just interested in learning more about psychology, neuroscience, and behaviour, plan to attend this special live lecture series. Although many colloquiums can seem intimidating and out of reach, these talks are made especially for you, the IntroPsych student. Each month a different faculty member will present a fascinating talk focusing on the most interesting and accessible research. If you can't make it in person, you can always watch the lecture later as it is posted on iTunes University. Please visit www.discoverpsychology.ca for details.

Tutorials: You will join a small tutorial section (capped at 26 students) led by a Teaching Assistant (TA) who is enrolled in or has completed PSYCH 3TT3: Applied Educational Psychology, a course designed specifically to help TAs lead effective tutorials and guide you through IntroPsych. Each year, our TAs receive top ratings from students across campus so don't be shy to ask questions. If you think your TA is especially amazing, consider nominating them for the Kathy Steele Award which honours the top TA of the year. Your TA will guide discussions, lead activities and demonstrations, and answer any questions you might have. Tutorials are updated each year by feedback from students and TAs.

Lobby Hours: Have a question or still confused about a specific concept? Need some one-on-one time? Want to meet TAs and students from outside your tutorial? Drop by the lobby of the psychology building (times posted on Avenue) to speak with TAs and students. If you have administrative questions please see Dr. Michelle Cadieux, the course coordinator, in PC 110. Her office hours are updated on Avenue weekly.

We have many talented and passionate members of the Instructional Staff and Development Team that work hard to bring you an outstanding course experience. IntroPsych was honoured with the 2010 President's Award for Excellence in Course and Resource Design. Our unique IntroPsych Program has been the topic of academic study and received widespread media attention in the Toronto Star, Globe and Mail, CHCH News, and Maclean's (not to mention Mac's own Daily News). Importantly, the continual development of the IntroPsych Blended Learning Environment model is supported by ongoing research. As the Director of the Applied Cognition in Education Lab, I am actively interested in teaching, learning and technology from both an academic and a practical perspective. For more information, visit http://www.science.mcmaster.ca/acelab/ or follow my twitter feed:@tlcjoekim

University can sometimes seem like an impersonal and strange place, especially for Level 1 students who are dealing with many adjustments. I hope that in exploring the course resources, you will not forget that there is a real live faculty member responsible for the IntroPsych Program—me! I have regular office hours (posted weekly on AVENUE) set aside solely to give me a chance to meet and talk with you. If you have a question, comment, complaint, concern, or just want to see and chat with a live faculty member, do come. Many students are reluctant to talk to a faculty member outside of class. Don't give in to the feeling! I have had many great conversations with students that have started off with a supposedly "silly" question.

As a Teaching Professor, my primary responsibilities are teaching and interacting with students; even my area of research interest concerns pedagogy - the formal study of teaching and learning. My goal is to help you understand and appreciate some of the really interesting and important things that we know (or think we know) about human thought and behaviour. In most fields - and as you will see, certainly in psychology - the simplest questions are often the most important and difficult to answer.

One last piece of advice—get involved in the course! IntroPsych is a fascinating world waiting to be explored by you! Keep up with the web modules, actively participate in tutorials, join the discussion forums, and attend the monthly Discover Psychology talks on Fridays. It really will make all the difference. Following each web module, I would also encourage you to participate in the feedback surveys. Many of the most popular interactive features were suggested by students just like you. I really do read every single comment and they have contributed enormously to minor and major changes made to all aspects of the course, and this includes the very course handbook you hold in your hands.

On behalf of all the wonderful people that work hard on the frontlines and behind the scenes, best of luck and have a great year!

Dr. Joe Kim

Web Modules: Interactive and easy to navigate

Navigation: You can pause, skip, and review each web module whenever you want. Tip: You can use the spacebar to quickly toggle between pause/play.

Viewing Options: You can view your web Modules with on an **outline** of the subtopics, **thumbnails** of the slides, or view **notes** a full transcript of Dr. Kim's narration.

Search: The Modules are fully indexed and can be searched by key words.

Checkpoints: Throughout the Modules you will find checkpoint questions designed to assess your understanding.

Shout Wall: Leave your mark. Comment or leave an interesting link for the rest of the class to see!

Polls: Share your opinion on topical questions related to the Module.

Media+, Docs+: Watch profiles of featured Psychology Faculty members and their research as well as bonus videos and games.

Course Content Schedule for Psychology 1X03/1N03 – Fall 2014

The general schedule for this course content is given below. Any changes to this structure will be announced on AVENUE. It is your responsibility to keep up-to-date with any schedule changes.

Week of	Web Module	Chapter Reading	Notes
Sep 1			**No Tutorial**
Sept 8	Introduction/Levels of Analysis	1	AVE Quiz 1
Sept 15	Research Methods 1 and 2	2	AVE Quiz 2
Sep 22	Classical Conditioning	3 (sections 1-5)	AVE Quiz 3
Sept 29	Instrumental Conditioning	3 (sections 6-8)	AVE Quiz 4
Oct 6	Problem Solving & Intelligence	Journal article	AVE Quiz 5
Oct 13	Language Library Research	4	AVE Quiz 6 **Thanksgiving – No tutorials this week**
Oct 20	Categories & Concepts	Journal article	AVE Quiz 7
Oct 27	**Midterm Recess**	---	**No Quiz or Tutorial this week**
Nov 3	Attention	5 (sections 1-4)	AVE Quiz 8
Nov 10	Memory	5 (sections 5-9)	AVE Quiz 9
Nov 17	Forming Impressions	6 (sections 1-2)	AVE Quiz 10
Nov 24	Influence of Others 1 and 2	6 (sections 3-7)	AVE Quiz 11
Dec 1	---	---	**No Tutorial**

WEEK of Sep 8: LEVELS OF ANALYSES

"It's not so much the pharmacology but the psychology that people overlook."
- Terry Robinson, prominent drug addiction researcher

If you were interested in understanding why some people become addicted to drugs, where would you begin? This was my challenge as I began my graduate career as an experimental psychologist. A first step would be to operationally define the problem. For me, this was clearly a behavioural phenomenon—an addict engages in persistent drug-seeking behaviour despite its negative consequences. If we could determine the conditions that reinforced this behaviour, perhaps we could train an addict to control his urges and prevent relapse. Our research provided compelling evidence that environmental cues are an important factor in relapse. This can explain why a recovering alcoholic walking by a familiar bar or a heroin addict holding an empty syringe may feel an almost irresistible urge to relapse. For the next several years I focused on understanding these interesting behavioural aspects of drug addiction. At the other end of the spectrum, many scientists considered drug addiction to be a purely pharmacological phenomenon to be studied at the cellular level. We were studying the same problem but asking very different questions. When I presented my research at conferences that focused on pharmacology, many researchers were very surprised to learn that non-pharmacological treatment can have such dramatic effects on drug action! In turn, I became increasingly interested in understanding the cellular and neural mechanisms of the observed behaviours. This broad experience helped me to recognize the importance of multiple levels of analysis to understand a problem as complex as drug addiction. Each perspective asks different questions leading to a richer understanding of the problem.

Kim, J.A., Bartlett, S., He, L., Nielsen, C.K., Chang, A.M., Kharazia, V., Waldhoer, M., Ou, C.J., Taylor, S., Ferwerda, M., Cado, D., & Whistler, J.L. (2008). Morphine-induced receptor endocytosis in a novel knockin mouse reduces tolerance and dependence. *Current Biology, 18*(2), 129-135.

Kim, J.A., Siegel, S., & Patenall, V.R. (1999). Drug-onset cues as signals: intra-administration associations and tolerance. *Journal of Experimental Psychology: Animal Behavioural Processes, 25*(4), 491-504.

Weekly Checklist:
- ☐ **Web Modules to watch: Introduction/Levels of Analysis**
- ☐ **Readings: Chapter 1**
- ☐ **AVE Quiz 1**

Upper Year Courses:
If you enjoyed the content in this week's module, consider taking the following upper year courses:
- PNB 2XD3 Integrative PNB
- PNB 4B03 History of Psychology

Module 1: Levels of Analysis – Outline

Unit 1: What is Psychology?

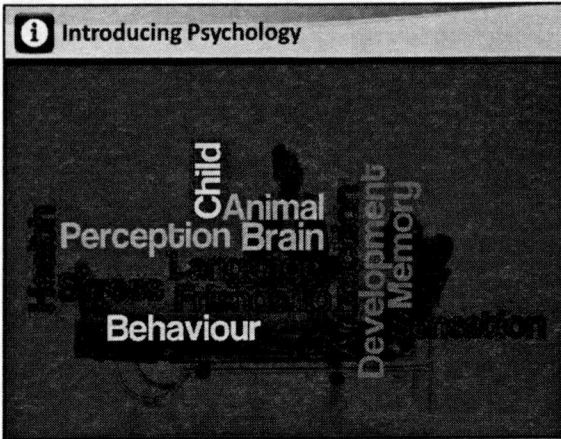

Introducing Psychology

The Study of Ourselves

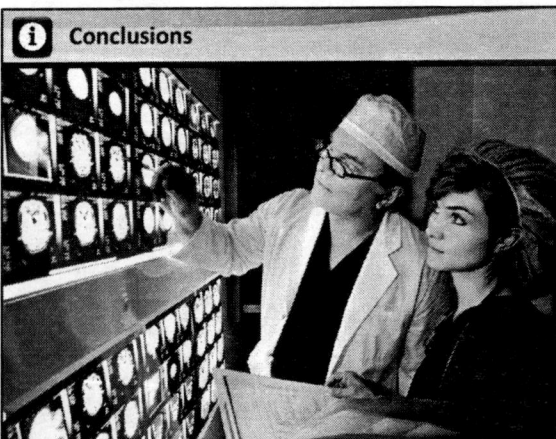

Conclusions

Psychologists work in many different settings and are employed by a variety of institutions.

_____
_____
_____
_____
_____
_____
_____
_____
_____
_____
_____

By systematically studying everyday phenomenon, we gain critical insight into our own behaviour.

_____
_____
_____
_____
_____
_____
_____
_____
_____
_____
_____
_____

Psychology hosts several schools of thought.

_____
_____
_____
_____
_____
_____
_____
_____
_____
_____
_____

Unit 2: A Brief History of Psychology

Psychology's Parents

Definition

Psy·chol·o·gy [sahy-**kol**-*uh*-jee]
The term psychology comes from the Greek words "psyche", which means **soul**.

Physiology's Influence

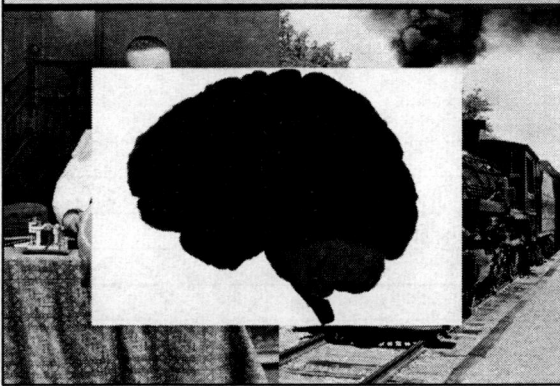

Psychology as an Independent Field

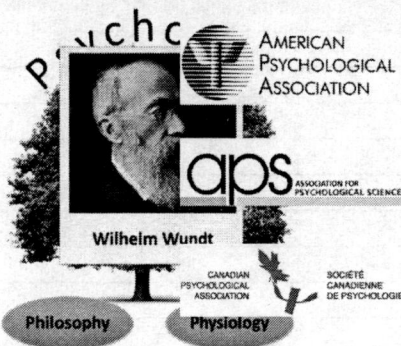

AMERICAN PSYCHOLOGICAL ASSOCIATION

aps ASSOCIATION FOR PSYCHOLOGICAL SCIENCE

Wilhelm Wundt

CANADIAN PSYCHOLOGICAL ASSOCIATION

SOCIÉTÉ CANADIENNE DE PSYCHOLOGIE

Philosophy   Physiology

_____
_____
_____
_____
_____
_____

The focus of psychological inquiry is heavily influenced by philosophy and physiology.

Ebbinghaus: _____
_____
_____

Aristotle: _____
_____
_____

Plato: _____
_____
_____

Descartes: _____
_____

Innovative tools allowed physiologists to research the transmission of nerve impulses.

Muller: _____
_____
_____

Helmholtz: _____
_____
_____

Flourens: _____
_____
_____

Psychology's emergence as an independent field was marked by several important events.

Wundt: _____
_____
_____
_____
_____

Hall: _____
_____
_____
_____
_____
_____
_____
_____
_____
_____
_____
_____
_____

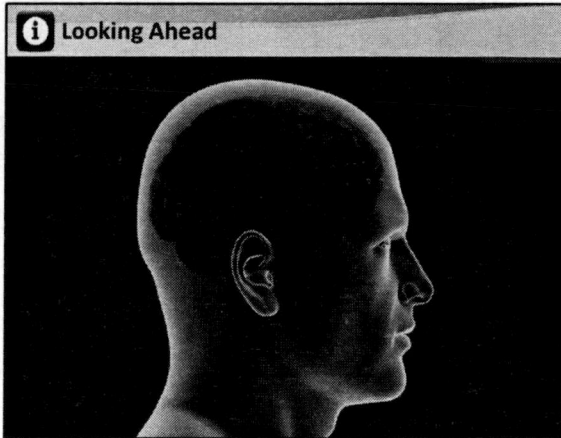
Looking Ahead

Today, psychology draws from a variety of fields to address several age-old questions.

_____
_____
_____
_____
_____
_____
_____
_____
_____

Unit 3: Introduction to Levels of Analysis

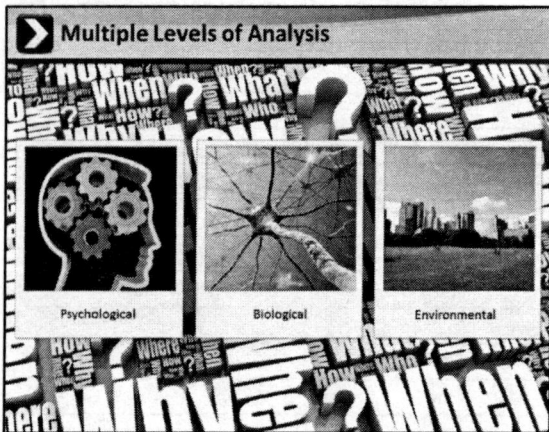
Multiple Levels of Analysis

Psychological   Biological   Environmental

Case Study: understanding depression using multiple levels of analysis.

Psychological: _____
_____
_____

Biological: _____
_____

Environmental: _____
_____
_____
_____

Unit 4: Introduction to Perspectives in Psychology

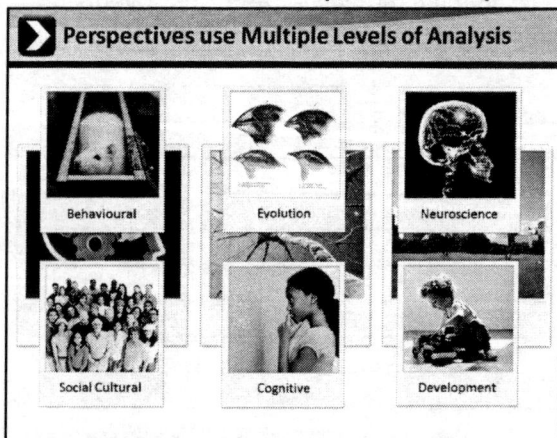
Perspectives use Multiple Levels of Analysis

Behavioural   Evolution   Neuroscience

Social Cultural   Cognitive   Development

Perspectives provide psychologists with a framework to conduct their research.

_____
_____
_____
_____
_____
_____
_____
_____
_____
_____
_____
_____
_____
_____

_____
_____
_____
_____

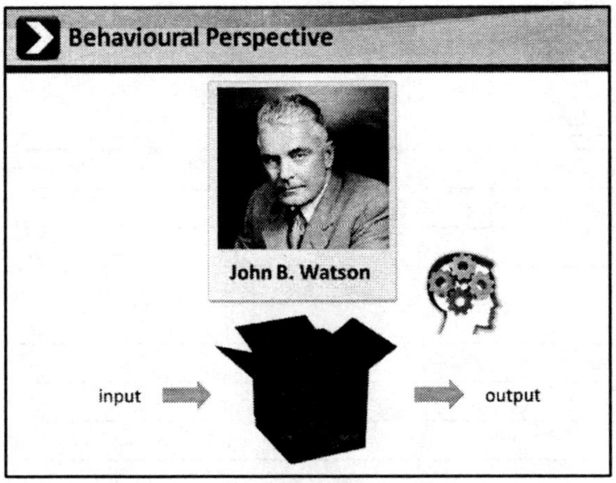

Behavioural Perspective

John B. Watson

input ➡ output

Behaviourists seek to analyze overt behaviour to answer psychological questions.

Watson: _____
_____
_____
_____
_____
_____
_____
_____
_____

Behavioural Perspective

YUCK!

BF Skin

Skinner: _____
_____
_____
_____
_____
_____
_____
_____
_____
_____
_____
_____
_____

Behavioural Perspective

_____
_____
_____
_____
_____
_____

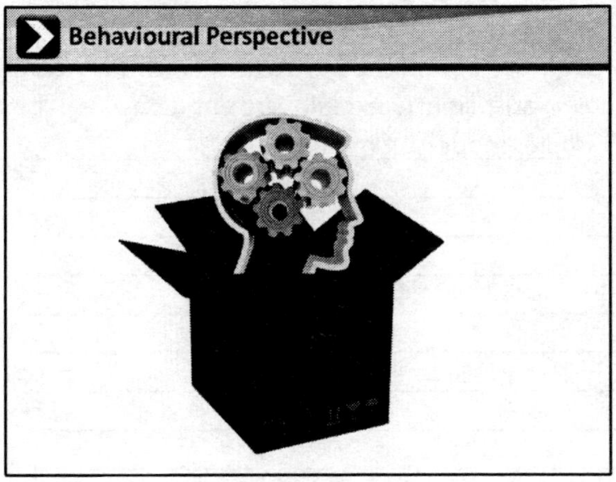

A completely behaviourist perspective is seldom adopted today.

_____
_____
_____
_____
_____
_____
_____
_____
_____
_____
_____
_____
_____
_____
_____
_____

Unit 5: Cognitive

The Cognitive Perspective

BF Skinner   John B. Watson

Cognitive psychologists argue that internal events must be considered in order to fully understand our behaviour.

_____
_____
_____
_____
_____
_____
_____
_____
_____

Aa Models in Cognition

Definition

Mod·els [mod-ls]
- Abstract representations of how the mind functions.
- Can be used to make predictions and design experiments.

Cognitive psychologists employ models to explain the functioning of complex systems.

_____
_____
_____
_____
_____
_____
_____
_____
_____
_____

Models in Cognition

New Model → Experiments → Adopt the new model

Revise or abandon new model ← Experiments ✗

Models are abandoned or revised until they can sufficiently explain current data.

Example: models of memory _____
_____
_____
_____
_____
_____
_____
_____
_____

_____
_____
_____
_____
_____
_____

_____
_____
_____
_____

Unit 6: The Biological Perspective and Reductionism

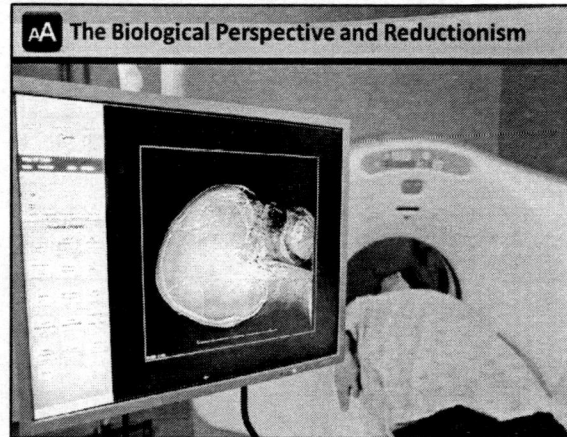
The Biological Perspective and Reductionism

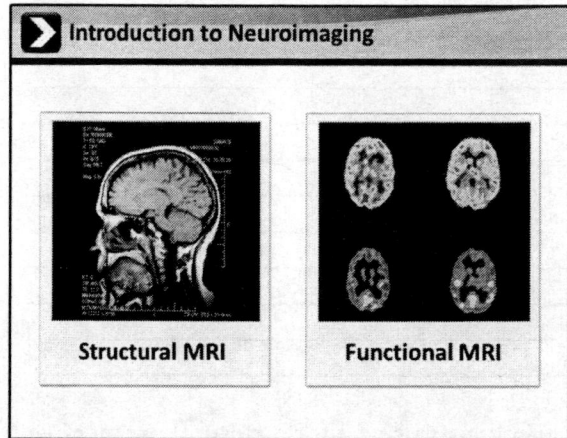
Introduction to Neuroimaging

Structural MRI   Functional MRI

Unit 7: Evolutionary and Developmental Perspectives

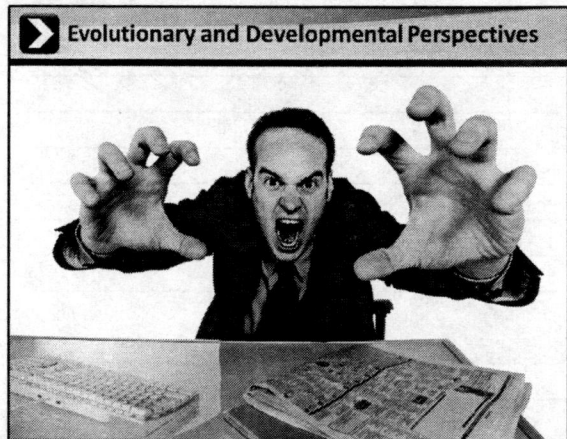
Evolutionary and Developmental Perspectives

_____
_____
_____

Psychologists prescribing to the biological perspective use brain mechanisms to explain behaviour.

Reductionism: _____
_____
_____
_____
_____
_____
_____
_____
_____
_____

Researchers use several imaging techniques to examine the brain.

Structural Neuroimaging: _____
_____
_____
_____

Functional Neuroimaging: _____
_____
_____
_____
_____
_____

Evolutionary psychologists seek to determine why certain behaviours exist in our society.

Proximate versus ultimate causes: _____
_____
_____
_____
_____
_____
_____
_____
_____
_____
_____
_____
_____

Evolutionary and Developmental Perspectives

Developmental psychologists examine behaviour over individual's lifespan.

Evolutionary versus Developmental: _____

_____
_____
_____
_____
_____
_____
_____
_____

Methods in Developmental Research

Developmental psychologists employ unique methods to conduct their research.

Habituation: _____

_____
_____
_____
_____
_____
_____
_____
_____
_____
_____

Unit 8: The Socio-cultural Perspective

The Socio-cultural Perspective

Psychologists using a socio-cultural perspective examine the effects that humans have on one another.

_____
_____
_____
_____
_____
_____
_____
_____
_____
_____
_____
_____
_____
_____

_____
_____
_____
_____
_____
_____
_____

Ethics

Milgram Experiment (1963)

Ethics guide academic research to ensure the wellbeing of all participants.

_____
_____
_____
_____
_____
_____
_____
_____
_____
_____

Unit 9: Perspectives on our Case Study

Understanding Depression – A Behavioural Perspective

What behaviours are associated with depression?

therapy   depressed brain   changed behaviour

A number of perspectives can help us understand depression.

Behaviourist perspective: _____
_____
_____
_____

Learned Helplessness: _____
_____
_____
_____
_____
_____
_____
_____
_____
_____
_____
_____
_____
_____
_____
_____
_____
_____
_____
_____
_____
_____

Understanding Depression – Other Perspectives

NATURE   or   NURTURE

_____
_____
_____
_____
_____
_____
_____

_____
_____
_____
_____
_____
_____

Module 1: Levels of Analysis – Courseware Exercise

Now that you've examined depression using a multidimensional approach, it is your turn to come up with an example and examine it using each perspective.

In choosing a topic, be sure to pick one that is general enough to be approached in many ways, but specific enough that you can find clear research on it. If you are having trouble coming up with an original example, think about the topics that you want to know more about or ones that drew you to this course. If you're REALLY stuck for a topic, try examining the topic of prejudice from the various perspectives.

Topic:_____

Next, you will be examining this topic using each perspective that was covered in the lecture. This will require a bit of research on your part. Using a search engine like PubMed, Google Scholar, or Web of Science, you can begin browsing through research articles and getting a sense of the breadth of your topic. Try to find lines of research, theories, or models that exemplify how you could study your topic through each perspective.

Learning:
Remember that learning involves relationships between cues and outcomes that lead to predicted responses. In taking this perspective, try to think of how your topic might be a learned behaviour or how it might affect learning in a specific way.

_____

_____

_____

Cognition:
Cognition is the study of the processing and organization that occurs in the internal mind in order to generate thoughts and behaviours. Think about how your topic might be related to specific memory or language principles or how the mind's subjective interpretation may be at play. This would be a great place to begin looking at cognitive models that exist for your topic.

_____

_____

_____

Neuroscience:
In neuroscience, a more physical perspective is taken to study the nervous system. Think about what areas of the brain are related to your topic and how they interact chemically via neurotransmitters and hormones. You may even examine the topic from a cellular level, if there is a specific subtype of neurons involved. However, be sure to understand that neuroscience is different from just pure physiology: neuroscience is interested in the nervous system, which is comprised of the brain, spinal cord, and peripheral nerves.

_____

_____

_____

Social Psychology:
Remember that social psychology is the study of how individuals influence one another in belief and behaviour. In looking at your topic using this perspective, guide your search by examining the influences that group and individual behaviours have on others.

_____

_____

_____

_____

Evolution:
Evolutionary psychology is the study of how behaviours evolve over many generations. The questions asked from this perspective are all-encompassing and seek to understand why certain behaviours evolved, while others did not. Many conforming behaviours have hypothesized evolutionary explanations that are supported by data. If your topic is related to abnormal behaviour, you may want to examine why it occurs in contrast to conforming behaviour and why it still occurs within the population.

_____

_____

_____

_____

Development:
Developmental psychologists examine and describe how behaviours change throughout a lifespan. In contrast to evolution, which examines why things change, development examines how things change. Remember though that development is a continual process and through this perspective we can examine changes early as well as late in life. Try to find out how your topic relates to variables such as aging and specific experiences in life.

_____

_____

_____

_____

You've now had valuable practice in thinking about a topic using multiple levels of analysis. This is an important way of thinking and will give you a strong perspective into psychology as your knowledge of the field grows. Although this course is organized by topic and level of analysis, it will be to your advantage to link ideas between lectures in order to build a multidimensional understanding of each concept.

Module 1: Levels of Analysis – Review Questions

1) Describe the biological approach.

    a. Compare and contrast the different types of neuroimaging. How are they useful?

2) Describe the basic approach of learning researchers.

    a. What makes behaviourism unique?

3) Why are ethics an important concern for social psychologists?

4) Compare and contrast the *Developmental* and *Evolutionary* approaches.

	Similarities	Differences
Developmental Approach		
Evolutionary Approach		

5) Match the following findings of different studies about human prejudice with the correct level of analysis.

"Favouring people like us and discriminating against those who are different from us have helped humans as a species to survive."

"There are a few areas of the brain that have been found to be associated with prejudice formation, but none of the findings have been conclusive."

"The results from this study provide new evidence that children who have been exposed to prejudice by their parents tend to be more discriminatory than their non-exposed counterparts."

"Prejudice can be explained as a form of heuristics, as it helps all of us to process incoming information quicker based on our earlier assumptions."

"Children understand that prejudice is not socially desirable, as they form the association between prejudice and negative consequences from early childhood.

Developmental Approach

Learning Approach

Neuroscience Approach

Evolutionary Approach

Cognitive Approach

Module 1: Levels of Analysis – Quiz Question

While sitting under an apple tree, Isaac Newton looked up and saw an apple falling towards him. After it quickly rolled away, he wondered what combination of brain areas was allowing him to perform such behaviour and why it would be advantageous for people to possess this behaviour. What two perspectives does Isaac use to address this issue?

      a. Social and learning (2.65%)
      b. Learning and developmental (23.44%)
      c. Developmental and neuroscience (26.09%)
      d. Neuroscience and evolution (47.18%)

This is an actual question from the 2011 Levels of Analysis Avenue Quiz that the majority of students struggled with. Fortunately, we are going to go through this question together to identify what sort of errors students were making and how to avoid them in the future. To do so, we will look at each answer option individually and identify why it is correct or incorrect.

To begin, it is important to identify the key information presented to you in the question stem. In this case, Mr. Newton performs a behaviours, and then asks two very important questions: 1) What combination of brain areas is responsible for this behaviour? 2) Why would it be advantageous for people to possess this behaviour? After picking out the key information let's now examine the answer options.

a. We need to figure out which two perspectives Mr. Newton is using when asking his questions. This option suggests that Newton is using the social and learning perspectives. Newton's first question involved the function of brain areas. According to the information presented in the web module, does the social perspective involve identifying the function of brain areas? No, thus it is not correct. Let's move to the next answer option.

b. Do either of the perspective in this option (learning/development) deal with functions of brain areas? No, it is not correct. Let's move to the next option.

c. Do either of the perspectives in this option (biological/development) deal with functions of brain areas? Yes! The biological perspective seeks to understand the physiological mechanisms of human thought and behaviour. Newton's second question dealt with why it would be advantageous for people to possess this dodging behaviour. Is this a question that might be asked by someone of the developmental perspective? Unlikely. The developmental perspective is, instead, concerned with how behaviour changes in one's lifetime. As a result this answer is incorrect. Let's move to our final option.

d. Do either of the perspectives in this option (biological/evolutionary) deal with functions of brain areas? Yes, the biological perspective! Now, does the evolutionary perspective deal with questions of why? Yes! The evolutionary perspective is concerned with the influence of genes and environment on behaviour over the history of a species. These psychologists would seek to understand why a behaviour evolved and what sorts of advantages or adaptive benefits the behaviour provided. (**Correct!**)

When completing the AVENUE tests, it is imperative that you actively pick out the critical information presented in the question stem. This will make it much easier to search for the correct answer option by comparing the information in the stem to that in the web module.

Key Terms

Behaviourist Perspective	Biological Perspective	Cognitive Perspective
Developmental Perspective	Evolutionary Perspective	Functional Neuroimaging
Models	Psychology	Reductionism
Socio-cultural Perspective	Structural Neuroimaging	

Module 1: Levels of Analysis – Bottleneck Concepts

Evolutionary vs. Developmental Perspectives
Behaviourism
Levels of Analysis

Evolutionary vs. Developmental Perspectives

People often get confused about these key concepts. Both perspectives deal with the development of behaviours over a period of time; the difference, however, lies in the timeframe they examine. The developmental perspective looks at the change of behaviours or traits over an individual's lifetime, while the evolutionary perspective looks at the change of behaviours or traits over many generations.

A developmental psychologist investigates how environmental and genetic factors influence and contribute to individuals developing traits such as problem-solving, sense of sight, or memory. They might ask questions that examine how brain changes during development influence children's ability to problem-solve or how nature and nurture contribute to the development of memory. On the other hand, the evolutionary perspective focuses on how and why behaviours and certain genetic traits are passed down from one generation to the next. In order for these behaviours or genes to appear in future generations, they must help the species by increasing the quality or quantity of their offspring. An evolutionary psychologist might ask why females in the animal kingdom are typically more selective when picking their mates. They would also explore why males benefit from mating with as many females as possible.

To make these concepts more concrete, let's take a look at the topic of Alzheimer's disease. A developmental psychologist would ponder whether a lack of mental stimulation influences an individual's chances of getting Alzheimer's disease and if there are any genetic factors that predispose individuals to Alzheimer's. The answer to both of these questions is yes. Mental stimulation has been shown to slow the onset of Alzheimer's disease and Alzheimer's seems to have a genetic component. An evolutionary psychologist, on the other hand, might examine why genes that predispose individuals to Alzheimer's disease exist in our population. One proposed answer is that Alzheimer's might represent a mechanism to slow down brain metabolism as humans age. This diminution in activity would have resulted in reduced energy requirements for aging individuals in an environment where food was scarce and a person's ability to gather food decreased dramatically as they became older. Unfortunately, this trait may have also resulted in Alzheimer's disease.

Remember, the key question when differentiating between the evolutionary and developmental perspectives involves asking whether the question being posed is one that spans many generations or just one lifespan.

Test your Understanding:

1) Which of the following statements does NOT appropriately analyze humans' love of sugary and fatty foods?

a) A developmental psychologist might ask how the amount of sugary foods parents give to their children affects how much sugary foods they eat when they grow older.
b) An evolutionary psychologist might say that sugary and fatty foods provide the most energy and so it made sense for people to develop a love for these foods in the past where food was often scarce.
c) A developmental psychologist might take at look at how the amount of physical activity someone does affects their tendency to eat sugary and fatty foods they eat.
d) An evolutionary psychologist might examine what genes become active when someone eats too many sugary and fatty foods and how these genes work over someone's lifetime.

Answer: Statement D. Options A and C are correct because a developmental psychologist investigates how certain traits develop over an individual's lifespan and factors that influence them (e.g. how parental care and physical activity affect a person's love for fatty foods). B is correct because an evolutionary psychologist looks at how traits develop over many generations; in this case, the psychologist is examining why preference for fatty foods would be selected over multiple generations. D is incorrect because a strictly evolutionary psychologist will not ask questions regarding the development of traits over an individual's lifespan.

2) Narcissistic personality disorder is a condition where an individual is obsessed with themselves and is convinced that they are superior to other people. What questions might you ask in the study of narcissistic personality disorder as a developmental or an evolutionary psychologist?

Answer: Answers could differ for this question. A developmental psychologist might ask how culture influences the development of the narcissistic personality. Interestingly, recent research has suggested that the increase in individualism in societies may contribute to the development of this personality disorder. An evolutionary psychologist might examine how being narcissistic helped our ancestors in their societies? One hypothesis is that thinking highly of yourself and being prideful may have helped obtain higher social status, allowing an individual better chances to reproduce. You can read more about this research in the following two articles:

1. Twenge, J.M., & Foster, J.D. (2010). Birth Cohort Increases in Narcissistic Personality Traits Among American College Students, 1982–2009, *Social Psychological and Personality Science, 1*(1), 99-106.

2. Cheng, J.T., Tracy, J.L., & Hendrich, J. (2010). Pride, personality, and the evolutionary foundations of human social status, *Evolution and Human Behavior, 31*(5), 334-347.

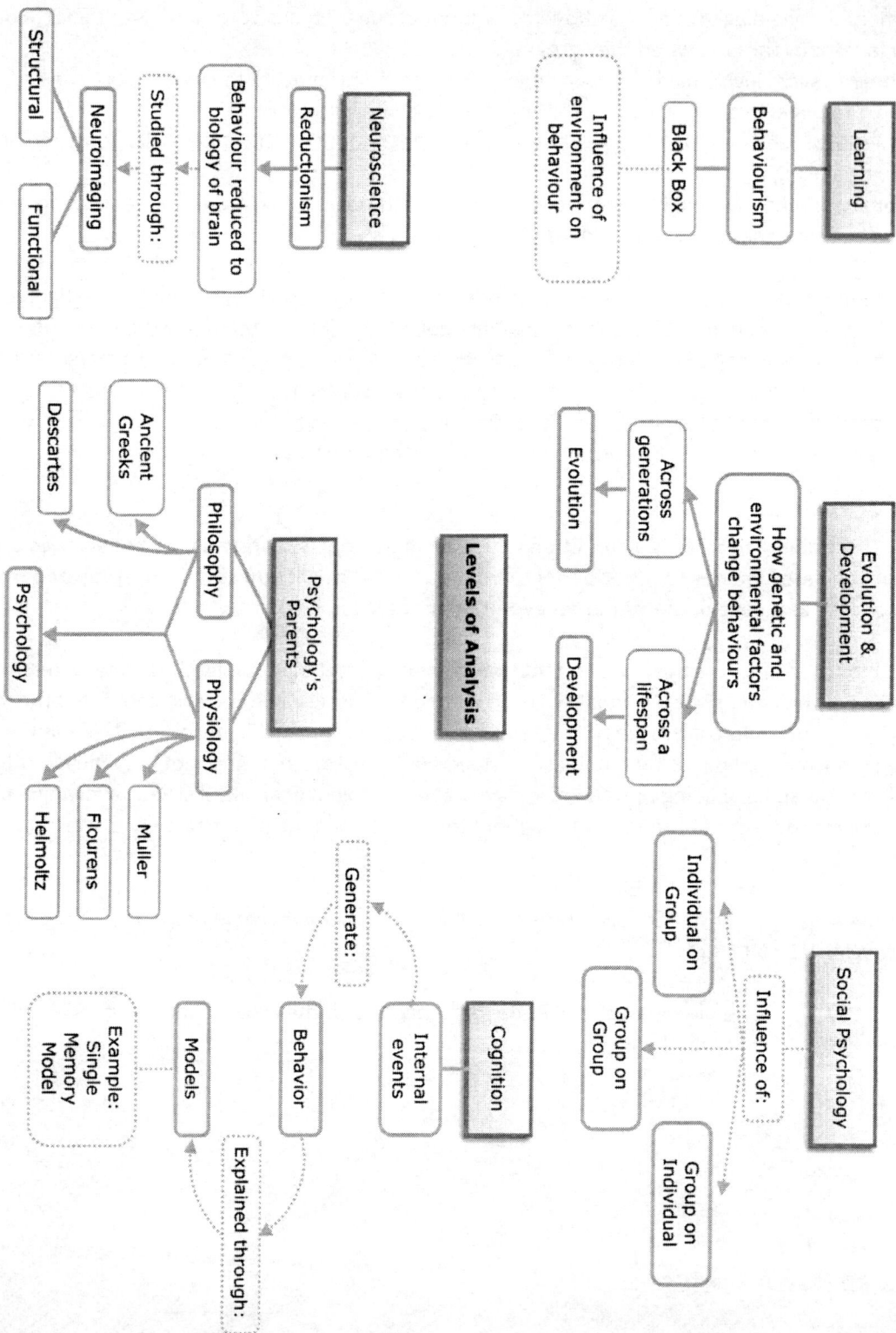

Levels of Analysis

Structural
Functional
Neuroimaging
Studied through:
Behaviour reduced to biology of brain
Reductionism
Neuroscience

Influence of environment on behaviour
Black Box
Behaviourism
Learning

Descartes
Ancient Greeks
Philosophy
Psychology's Parents
Psychology
Physiology
Helmoltz
Flourens
Muller

Evolution
Across generations
How genetic and environmental factors change behaviours
Evolution & Development
Development
Across a lifespan

Generate:
Behavior
Models
Example: Single Memory Model
Explained through:
Internal events
Cognition

Individual on Group
Group on Group
Influence of:
Social Psychology
Group on Individual

WEEK of Sep 15: RESEARCH METHODS

"Extraordinary claims demand extraordinary proof."
- Carl Sagan, noted astronomer and science journalist

There's an extraordinary dog named Jaytee who lives in Ramsbottom, England. Over the years, he has developed a special bond with his owner Pam. Pam's parents noticed that Pam always seemed to arrive home shortly after Jaytee went out to the front porch to wait for her. This was true even when Pam returned home at unusual times. Intriguingly, it seemed that Jaytee could sense when Pam was starting her journey home. This special ability did not escape the attention of the media and Jaytee was featured on several television shows that celebrated his paranormal powers. Things got even more interesting when a study supporting the claim was published by Dr. Rupert Sheldrake, a researcher interested in 'psychic pet' phenomena. However, you might want to consider alternate explanations. Can Jaytee pick up his owner's scent at great distances? Are Pam's parents unknowingly giving cues that Jaytee detects? Do correct responses by Jaytee coincide with Pam's normal routine? The psychic dog underwent a series of experiments under the direction of Dr. Richard Wiseman, a paranormal sceptic who was critical of Sheldrake's study. Over a series of experiments, several factors were controlled: Pam's return time home was randomly selected; residents of the home were kept unaware of Pam's return time; and, Pam drove home in a different car (to eliminate characteristic car sounds). Before the experiment began, Wiseman operationally defined psychic ability as "a trip to the porch preceding Pam's return home within 10 minutes." As with many sensational claims, it turned out that the simplest explanation was correct—under these controlled conditions Jaytee turned out to be a normal dog. There were certainly a lot of resources dedicated to test an admittedly flimsy claim. However, think about the news reports, advertisements, and urban myths you hear making extraordinary claims in the absence of solid scientific research. It is often easier to passively accept these statements than to be a healthy sceptic. More than likely, a miracle pill is not so miraculous, a fortune is not waiting for you to claim, and pet dogs often lack psychic powers.

Sheldrake, R., & Smart, P. (1998). A dog that seems to know when his owner is returning: Preliminary investigations. *Journal-Society for Psychical Research, 62*, 220-232.

Wiseman, R., Smith, M., & Milton, J. (1998). Can animals detect when their owners are returning home? An experimental test of the 'psychic pet' phenomenon. *British Journal of Psychology, 89*(3), 453-9.

Weekly Checklist:
- ☐ **Web Modules to watch: Research Methods 1 and 2**
- ☐ **Readings: Chapter 2**
- ☐ **AVE Quiz 2**

Upper Year Courses:
If you enjoyed the content in this week's module, consider taking the following upper year courses:
- PNB 2QQ3 Research Practicum
- PNB 2XE3 Descriptive Statistics
- PNB 2XF3 Research Methods
- PNB 3XE3 Inferential Statistics

Module 2: Research Methods I – Outline

Unit 1: Introduction to Scientific Research

The Scientific Method

The Scientific Method

1) Construct a theory

2) Generate hypothesis

3) Choose research method

4) Collect data

5) Analyze data

6) Report the findings

7) Revise existing theories

The Scientific Method

Hypothesis — Form a testable statement guided by theories that make specific predictions about the relationship between variables.

Theory — Collect a general set of ideas about the way the world works.

The Scientific Method

Analyze Data — Understand the data and discover trends or relationships between the variables.

Collect Data — Take measurements of the outcomes of the test.

Research Method — Determine the way in which the hypothesis will be tested.

Hypothesis

Theory

The scientific method provides a framework within which scientists conduct their research.

_____

_____

_____

_____

_____

_____

_____

_____

_____

_____

Theory: _____

_____

_____

_____

_____

Hypothesis: _____

_____

_____

_____

_____

_____

Research method: _____

_____

_____

Collect data: _____

_____

_____

Analyze data: _____

_____

_____

_____

The Scientific Method

Revise Theories — Incorporate new information into our understanding of the world.

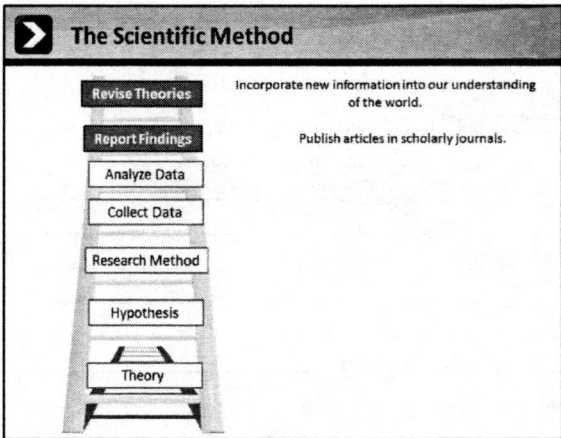

Report Findings — Publish articles in scholarly journals.

Analyze Data

Collect Data

Research Method

Hypothesis

Theory

Report findings: _____
_____
_____
_____
_____

Revise theories: _____
_____
_____
_____
_____
_____

Unit 2: Conducting an Experiment

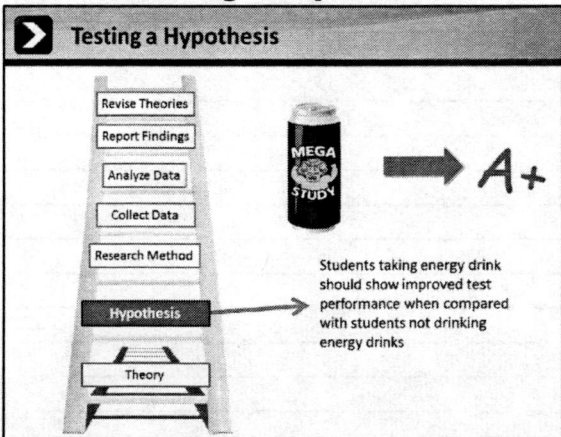

Testing a Hypothesis

Revise Theories

Report Findings

Analyze Data

Collect Data

Research Method

Hypothesis

Theory

MEGA STUDY → A+

Students taking energy drink should show improved test performance when compared with students not drinking energy drinks

Testing a Hypothesis

Single experience might not be representative.

Personal experience might not represent others.

Cannot be sure that result is due to energy drinks alone.

Case Study: Eric Tests the Mega Study Energy Drink

Eric's theory: _____
_____
_____

Eric's hypothesis: _____
_____
_____

Anecdotal evidence: _____
_____
_____
_____

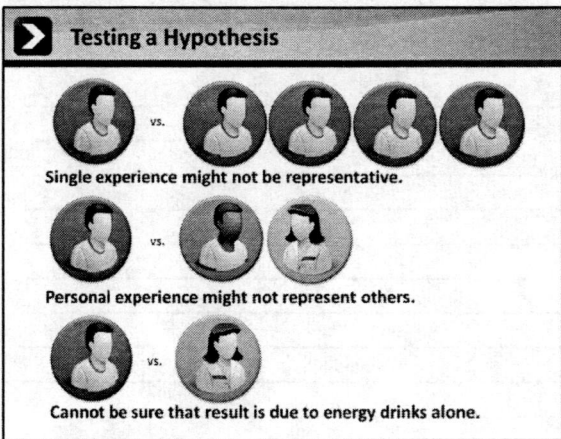

Without rigorous testing, it can be difficult to establish a cause and effect relationship.

_____
_____
_____
_____
_____
_____
_____
_____
_____
_____

_____
_____
_____
_____
_____

Using an Experiment

Definition

Ex·per·i·ment
[ek-**sper**-uh-ment]
- Scientific tool used to measure the effect of one variable on another

In·de·pend·ent Var·i·a·ble
[in-di-**pen**-duhnt] [**vair**-ee-uh-buhl]
- Variable manipulated by the scientist

De·pend·ent Var·i·a·ble
[dih-**pen**-duhnt][**vair**-ee-uh-buhl]
- Variable being observed by the scientist

Using an Experiment

Gets good grade → Drink is successful

Gets bad grade → Drink is failure

Independent variable not properly manipulated!

Unit 3: Control Groups

Using Control Groups

Differ only in independent variable

A+ C-

Difference must be due to independent variable

An experiment is one way with which scientists can begin to tackle complex research questions.

Experiment: _____

_____

_____

Independent variable: _____

_____

Dependent variable: _____

_____

_____

Eric's initial experimental methods contained several flaws.

_____

_____

_____

_____

_____

_____

_____

_____

_____

_____

_____

_____

_____

Control groups must be carefully selected to conduct an effective experiment.

_____

_____

_____

_____

_____

_____

_____

_____

_____

_____

_____

_____

_____

_____

_____

_____

Within Subjects Design

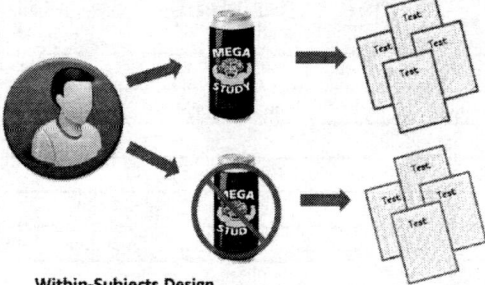

Within-Subjects Design
- Manipulating the independent variable within each participant to minimize the effect of external variables on the dependent measure

Subject Designs

Practice Effect
- Improved performance over the course of an experiment due to becoming more experienced

Between Subjects Design

Between Subjects Design
- One group acts as the control group

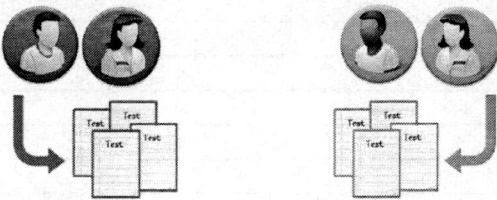

Confounding Variable
- A variable other than the independent variable that has an effect on the results

A within subjects design allows experimenters to discount the effects of differences between groups.

_____
_____
_____
_____
_____
_____
_____
_____

Nevertheless, a within subjects design is prone to practice effects.

_____
_____
_____
_____
_____
_____
_____
_____

A between subjects design compares the independent variable's effect on two groups.

_____
_____
_____
_____
_____
_____
_____
_____
_____

Unit 4: Sampling

Selecting Subjects

Results from very specific groups of participants CANNOT be generalized to other groups

High school average = 93%

Selecting Subjects

Population
All undergraduate students

Random Sample
■ Choosing a sample at random from the entire population

Selecting Subjects

Random Assignment
■ Assigning subjects to either the experimental or control group at random to avoid any biases that may cause differences between the groups of subjects

_____
_____
_____
_____
_____
_____

Participants must be carefully selected to allow researchers to accomplish the goals of their experiment.

_____
_____
_____
_____
_____
_____
_____
_____
_____
_____

Random sampling is often used to ensure that participants accurately represent the population of interest.

_____
_____
_____
_____
_____
_____
_____
_____
_____

Random assignment further prevents systematic differences between the two groups.

_____
_____
_____
_____
_____
_____
_____
_____
_____
_____
_____
_____
_____
_____
_____
_____

Unit 5: Conducting an Experiment

Subject Biases

Placebo Effect
- Effect that occurs when an individual exhibits a response to a treatment that has no related therapeutic effect

Subject Biases

Experimental Group

Control Group

Blinding
- When participants do not know whether they belong to the experimental or control group, or which treatment they are receiving

Experimenter Biases

Test improvement may be due to an experimenter's influence, not the experimental manipulation

Experimenter Bias
- Actions made by the experimenter, intentionally or not, to promote the result they hope to achieve

_____
_____
_____
_____
_____
_____

Placebo effects can have an undesirable influence on the experimental results.

_____
_____
_____
_____
_____
_____
_____
_____
_____
_____

Researchers have devised several clever methods to remove unwanted effects on their experiments.

Subject bias: _____
_____
_____
_____

Blinding: _____
_____
_____
_____

Experimenter bias: _____
_____
_____
_____

Double-blind studies: _____
_____
_____
_____
_____
_____
_____
_____
_____
_____

Module 2: Research Methods I – Courseware Exercise

Now that you have a grasp on how to correctly design a scientific experiment, take a look at the following experiment and critique it to **find at least 5 problems** in the way that it is designed.

<u>*Lighting in Factories 1*</u>
The aim of this study is to determine conditions that lead to more efficient work environments. AMC, a factory, has volunteered to be the subject of our experiment. Every week, a condition will be changed, and performance of workers will be observed.

1) _____
2) _____
3) _____
4) _____
5) _____

Now take a look at the following corrected experimental design and find **at least 5 ways in which it improves** on the previous experimental design.

<u>*Lighting in Factories 2*</u>
The aim of this study is to determine whether brighter lighting increases production in factories that produce shoes. Forty factories in China, Mexico, and Canada have been randomly selected for study. Half of these factories will receive brighter lighting through the installation of desk lamps, while the other half of the factories will receive no manipulations. Differences between the average output of the two groups before the experiment are negligible. After two weeks, output numbers will be collected and a t-test will be performed between the control group and the experimental group.

1) _____
2) _____
3) _____
4) _____
5) _____

It is important to note that this second experiment is not perfect either! It is important to always be critical of any experimental method used in science in order to constantly improve and reduce errors and biases. As an added challenge, try to find two problems in the second experimental design, as well as solutions to solve each of these.

1) Problem: _____

   Solution: _____

2) Problem: _____

   Solution: _____

Module 2: Research Methods I – Review Questions

1) Identify and describe the steps of the scientific method, with reference to the following social psychology experiment:

   Middlemist, R. D., Knowles, E. S., & Matter, C. F. (1976). Personal space invasions in the lavatory: suggestive evidence for arousal. *Journal of personality and Social Psychology, 33*(5), 541-546.

   Step 1: Access Google Scholar at: http://scholar.google.ca
   Step 2: Type "Urinal Experiment" into the search box
   Step 3: Locate the article entitled "Personal space invasions in the lavatory: suggestive evidence for arousal"

2) Describe the difference between a theory and a hypothesis.

3) Refer to the scenario below for questions a through d.

   Dr. Burns is conducting an experiment to study the effect of eating breakfast on university students' ability to learn in lectures. In condition A, all students have breakfast prior to watching their psychology module. In condition B, the students are instructed to skip breakfast, and eat an early lunch after watching the module. One week later, all of the students are tested on the material learned in the morning lectures.

   a. What is the independent variable? What is the dependent variable?
   b. Is this a within-subjects or between-subjects design? Justify your answer. How could experiment be manipulated to make the it the other type of participant design?
   c. Why is random sampling necessary for a quality experiment?
   d. Referring to the above scenario, which of the following variable is **NOT** a potential confounding variable?
      i. The time that students eat lunch after watching the module
      ii. The amount of sleep students have had the day prior to the testing
      iii. The amount of time students take to review the module material
      iv. The location where students watched the module

4) Which of the following experiments is a double blind study?
   a. To test the effects of a new psychotherapy technique, a clinical psychologist provides the new psychotherapy treatment for half of his clients, and provides the original treatment for his other clients. A month later, he evaluates all of his clients' progress.
   b. A neuropsychologist is conducting functional MRI studies to examine areas of the brain associated with humour. She asks a colleague to randomly assign half of her subject to be presented with a comedic sitcom while having the MRI scan and the other subjects to be presented with the weather forecast while having the scan. She then interprets the MRI images with the assistance of her colleague.
   c. A social psychologist is studying how children handle conflict in the playground. He asked his research assistant to explain to half to children that they will be exposed to a bully and have to behave aggressively, and the other half of the children that they will have to behave passively. Not knowing which group is which condition, he observes the children on how they handle the conflict.
   d. To study the reaction time of toddlers, a psychology student asks a friend to present a group of 2 year old children and another group of 3 year olds with a toy, while the student is out of the room. All the children are required to press a red button as soon as they see the toy. The student then interprets the data on her own to determine which group has a faster reaction time.

Module 2: Research Methods II – Outline

Unit 1: Introdution to Scientific Research

Types of Descriptive Statistics

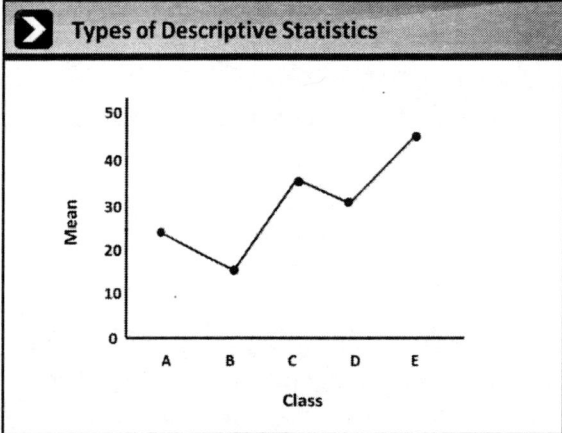

Creating a Histogram

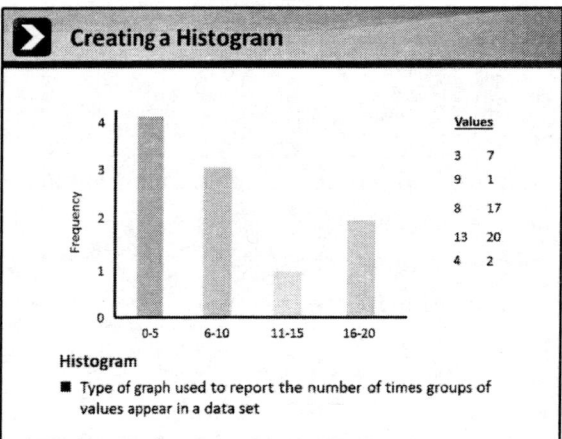

Histogram
- Type of graph used to report the number of times groups of values appear in a data set

Frequency Distributions

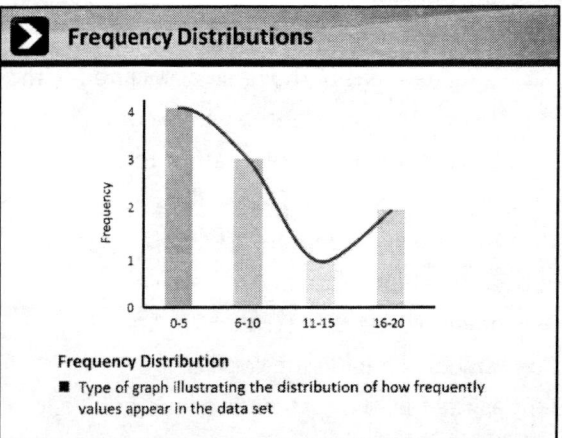

Frequency Distribution
- Type of graph illustrating the distribution of how frequently values appear in the data set

Raw data from an experiment can be converted into many easy-to-understand formats.

Descriptive statistics: _____

_____

_____

_____

_____

_____

_____

_____

_____

_____

_____

Histogram: _____

_____

_____

_____

_____

_____

_____

_____

_____

_____

_____

Frequency distribution: _____

_____

_____

_____

_____

_____

_____

_____

_____

_____

_____

_____

_____

Normal Distribution

Normal Distribution
- A distribution with a characteristic smooth, symmetrical, bell-shaped curve containing a single peak

Measures of Central Tendency

Mean
- The average value of a data set

$$\bar{x} = \frac{1}{n} \cdot \sum_{i=1}^{n} x_i$$

Percentage Grades:	Percentage Mean
89 90	
90 91	Mean = $\frac{89 + 90 + 90 + 91 + 93}{5}$
93	= 90.6

The Median

Median
- The centre value in a data set when the set is arranged numerically

---

A normal distribution is a special kind of distribution.

_____
_____
_____
_____
_____
_____
_____
_____
_____

Measures of central tendency allow researchers   to describe their data.

Mean: _____
_____
_____
_____

The effect of outliers: _____
_____
_____
_____

Median: _____
_____
_____
_____
_____
_____
_____
_____
_____
_____
_____
_____
_____

The Mode

Mode: _____
_____
_____
_____
_____

Disadvantages of exclusively using measures of central tendency: _____
_____
_____
_____
_____

Standard Deviation

Smaller Spread = Smaller Standard Deviation

Mean

Larger Spread = Larger Standard Deviation

Measures of variability: _____
_____
_____
_____
_____
_____

Standard deviation: _____
_____
_____
_____
_____

Unit 2: Inferential Statistics

Inferential Statistics

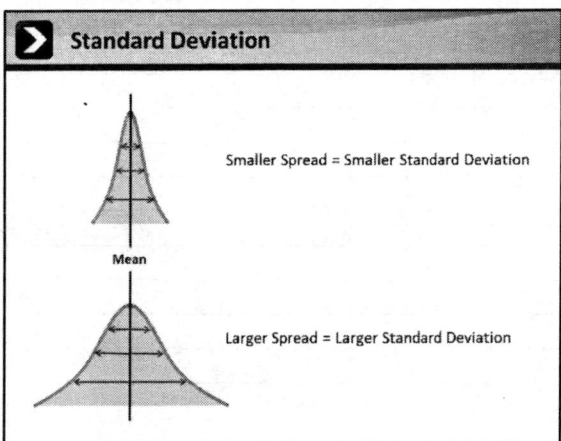

Inferential Statistics
- Statistics that allow us to use results from samples to make inferences about overall, underlying populations

Inferential statistics allow researchers to make inferences about their population of interest.

_____
_____
_____
_____
_____
_____
_____
_____
_____
_____
_____
_____
_____

_____
_____
_____
_____
_____
_____

Hypothesis Testing

Which distribution does this participant belong to?

Energy Drink has no effect – both groups belong to the same population

Energy drink has positive effect – experimental group belongs to different population

T-Test and P-Value

T-Test
- A statistical test that considers each data point from both groups to calculate the probability that two samples were drawn from the same population

DATA SET 1

DATA SET 2

the Magical "T-test" Machine

P-Value

P-Value
- A value expressing the probability calculated by the t-test

Statistical Significance: Setting a Criterion

| p = 0.10 | p = 0.20 | p = 0.07 |
| p = 0.08 | p = 0.11 | p = 0.21 |

Not significant. Greater than 5% probability of obtaining the data by chance.

| p = 0.045 | p = 0.001 | p = 0.01 |
| p = 0.02 | p = 0.03 | p = 0.05 |

Significant. Less than 5% probability of obtaining the data by chance.

A t-test is a method of applying inferential statistics to your data set.

Statistical significance:

Unit 3: Reviewing Experimental Design

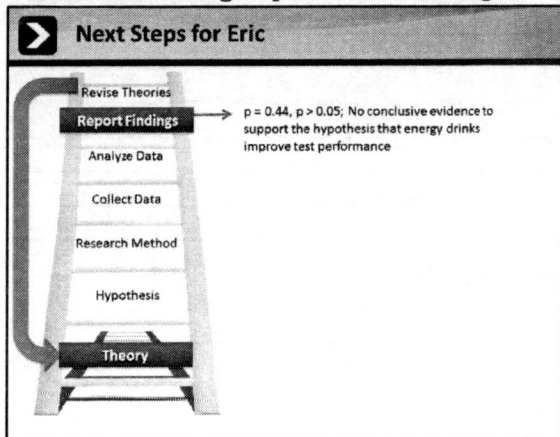

Next Steps for Eric

Revise Theories

Report Findings → p = 0.44, p > 0.05; No conclusive evidence to support the hypothesis that energy drinks improve test performance

Analyze Data

Collect Data

Research Method

Hypothesis

Theory

Unit 4: Observational Research

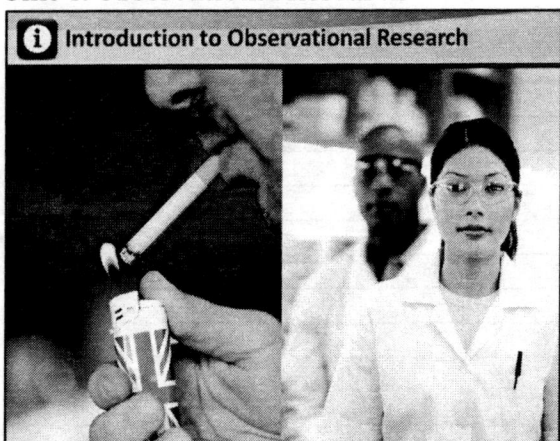

Introduction to Observational Research

Observational research can be an important tool for scientists in certain situations.

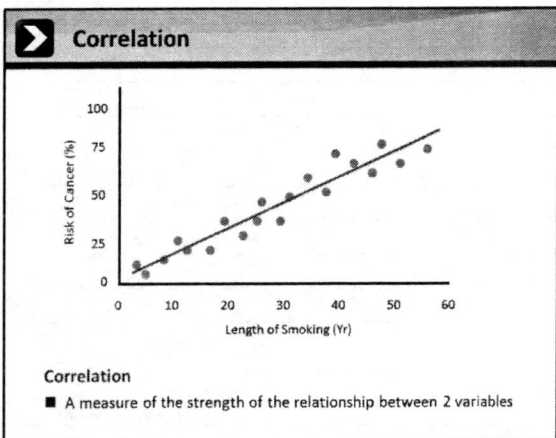

Correlation

Correlation
- A measure of the strength of the relationship between 2 variables

The relationship between two variables can be described by means of a correlation.

Measuring Correlation

r = 1
Perfect positive correlation

r = -1
Perfect negative correlation

Measuring Correlation

r = +1

r = 0

r = -1

r = +0.5

r = -0.5

The correlation coefficient provides information about the strength and direction of the relationship between two variables.

Correlation Does Not Equal Causation

Correlations must not be used to determine a cause and effect relationship.

Module 2: Research Methods II – Courseware Exercise

On a recent test, Professor A's class had a fantastic average of 85%. Professor A, being a bit of a show-off, came gloating about his far superior average to Professor B, whose class had an average of 79%. However, upon examination of the class distribution, Professor B realized that Professor A may have been incorrect to assume that there was such a large difference between the two classes.

Here are the two distributions:

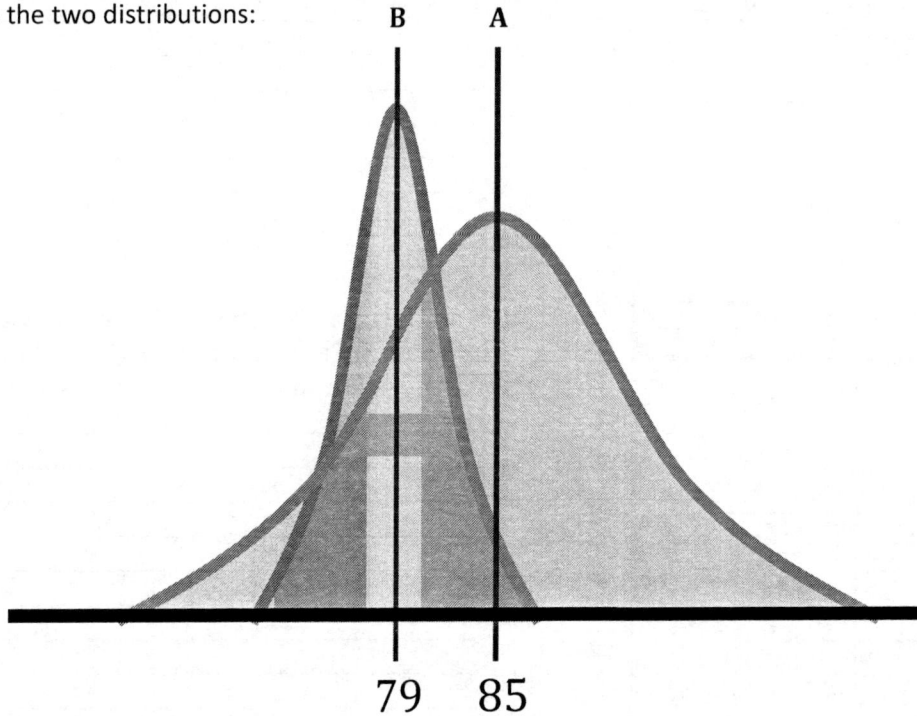

79   85

Using the graphic representation of the two distributions, can you think of one disadvantage of the spread of Professor A's class scores? Assuming that all else is equal, what could this inform you about how well students learn from Professor A?

_____

_____

_____

_____

_____

Can you think of a way in which further statistical analysis may tell us more about how these distributions compare? What are the two possible outcomes of this statistical analysis?

_____

_____

_____

_____

Module 2: Research Methods II – Review Questions

1) Using the following data set of scores, construct a histogram.

2	2	4	5	6	6	6	6	9	10

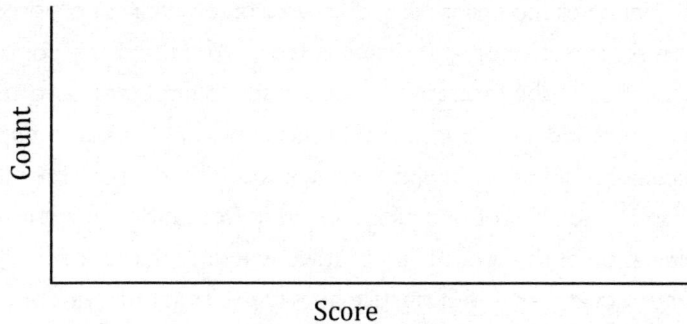

Score

2) Based on the above data set, what is the Mean? Median? Mode?

Mean: _____   Median: _____   Mode: _____

3) What are the advantages and disadvantages of the three measures of central tendency?

	Advantage	Disadvantage
Mean		
Median		
Mode		

4) Which of the following curves has the greatest standard deviation? The smallest standard deviation?

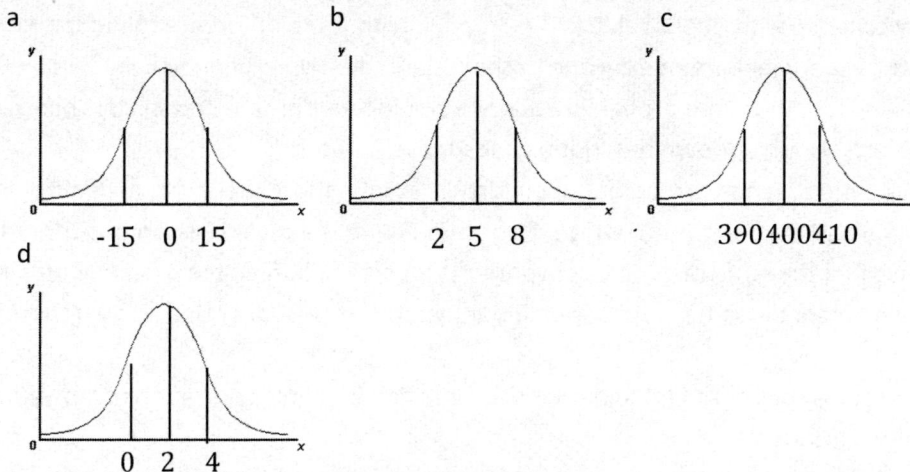

a   -15 0 15
b   2 5 8
c   390 400 410
d   0 2 4

5) Define the concepts "sample" and "population". Would an experimenter prefer to conduct an experiment on a population or sample? Why? Explain the importance of samples in inferential statistics.

6) What do we mean when we say a result is statistically significant?

7) What are correlations? Why must we be careful when measuring correlations? Consider the example below. "*Men who are obese may have less brainpower than their trim counterparts, according the research done at University of Boston. The study found that men with a BMI of 30 or more scored on average **23%** lower in tests of mental acuity. However, the study did not make clear whether the obesity was caused by the lower IQ or vice versa.*" Imagine the implications of this study if causation was implied.

Module 2: Research Methods I & II – Quiz Question

Which of the following best characterizes the meaning of the **p-value**?

     a. The result of a calculation of the probability that we would find a set of results by chance if our hypothesis was **correct** and all of our participants were from the **same population**. (24.39%)

     b. The result of a calculation of the probability that we would find a set of results by chance if our hypothesis was **incorrect** and all of our participants were from the **same population**. (40.18%)

     c. The result of a calculation of the probability that we would find a set of results by chance if our hypothesis was **correct** and all of our participants were from **different populations**. (20.53%)

     d. The result of a calculation of the probability that we would find a set of results by chance if our hypothesis was **incorrect** and all of our participants were from **different populations**. (14.91%)

This is an actual question from the 2011 Research Methods Avenue Quiz that the majority of students struggled with. Again, we are going to go through this question together to identify what sort of errors students were making and how to avoid them in the future. To do so, we will look at each answer option individually and identify why it is correct or incorrect.

Before we begin to tackle this question, it is important that you actively attempt to understand the material presented in the web module. This question requires a solid understanding of the p-value concepts presented in the web module and serves as an example of how difficult the AVENUE tests can be if you only have a superficial understanding of the content. Explaining a concept to yourself in your own words is an effective alternative to simply scanning and rereading your notes. Try doing this with the p-value now. Come up with a description of what the p-value is and compare your description to those in the answer options. Here is an example of the description that you may come up with: the p-value is the probability that the difference found between two groups or the results of the experiment are merely due to chance and the two groups actually belong to the same greater population. Indeed, this is the probability that we would find the same results by chance even if our hypothesis was incorrect. Using your own description, let's move forward.

     a. At first glance this option seems correct, but upon further examination, there is a small difference between this option and the description above. This option defines a p-value as, among other things, the probability that we find the same result if our hypothesis is correct. In fact, the p-value would be the probability that we obtain the same results even if our hypothesis was incorrect. Scratch this option—it is incorrect.

     b. Again, this option seems correct at first and seems to be in line with my description of a p-value. Let's hold onto this option. (**Correct!**)

     c. This option is talking about the probability that we find the same result if our hypothesis is correct. Scrap this option as well.

     d. Ah, this option seems correct as well. It correctly defines a p-value as the probability of achieving the same result by chance even if the hypothesis is incorrect. However, it ends with a clause stating, "if the two groups are from different populations". This is opposite to my description.

After considering all the answer options, we are left with only **option B**. Remember, it is important to be able to understand the information in the web modules to the extent that you are able to explain and describe the content in your own words.

Key Terms

Between-Subjects Design	Hypothesis	Population
Blinding	Independent Variable	Practice Effect
Causation	Inferential Statistics	Random Assortment
Control Group	Mean	Research Method
Correlation	Measures of Central Tendency	Sampling
Correlation coefficient (r)	Measures of Variability	Scientific Method
Data	Median	Scatter-plot
Descriptive Statistics	Mode	Standard Deviation
Dependent Variable	Normal Distribution	Statistical Significance
Double-Blind Study	Observational Studies	Subject/Participant
Experimental Group	Outliers	T-test
Experimenter Bias	P-value	Theory
Frequency distribution	Participant Bias	Within-subjects Design
Histogram	Placebo Effect	

Module 2: Research Methods I & II – Bottleneck Concepts

T-test vs. P-value
Single Blind vs. Double Blind
Causation vs. Correlation
Independent vs. Dependent Variable

T-test and p-value

When running an experiment, a researcher only tests a sample of the population. This is done to avoid spending excessive time and resources testing the entire population. An independent variable is manipulated and the investigators look for differences in the participants or their performance depending on the manipulation. However, to make their research relevant, experimenters must also ensure that the findings of their experiment reflect the general population. One way to do this is through a t-test. A t-test determines whether the manipulation of the independent variable had an effect or if the results could have been just because of chance.

As an example, pretend that the marks in a math class were 60, 65, 70, 70, 70, 75, 75, 75, 80, 85, 90, and 95. If you randomly split the marks into two groups, you could potentially end up with one group having students with the marks: 60, 65, 70, 70, 75 and 80, and the other group with 70, 70, 75, 75, 90, and 95. Their groups' average mark would be 70 and 79 respectively. However, did you notice that you did not do anything to manipulate students' marks and you found this difference simply by random chance? This same scenario could happen when a scientist runs an experiment. By chance, they could see a difference between their control and experimental groups! Running a t-test gives researchers a probability, known as the p-value, which states the likelihood that the experimenters obtained their results by chance. If the p-value is less than 0.05, then there is less than a 5% chance that the scientists obtained their results by luck and thus, greater than 95% chance that their manipulation of the variable caused the difference between the two groups. A p-value less than 0.05 is generally regarded as a significant result where the difference between the groups is likely due to the manipulation of the independent variable. (Note: You do not have to know the details of how to actually perform a t-test; you just need to know the meaning of the results.)

Test your Understanding

Brian wants to know if loud noises affect how well a person remembers. He gives a list of 50 words to 2 groups of 20 randomly selected people. They are to remember as many words as possible in 10 minutes. One of the groups is placed in a silent room while the other is placed in a room where construction noises are being played on loudspeakers. Brian then tests to see how many words each person remembered. He runs a t-test and is 90% confident that the noise reduced the people's ability to memorize words.

a) What p-value did he obtain? Are the results considered significant?

Answer: He obtained a p-value of 0.10. If he is only 90% confident that there is a true difference between the groups, it means there is a 10% chance he could have gotten the results by chance. This means that his results are not significant because his p-value is > 0.05.

b) Brian suspects that there is an actual difference between the groups but the data just haven't reached significance yet. What can he do in order to obtain significant results?

Answer: Brian may obtain significant results if he performed the experiment with more participants. With more participants, Brian diminishes the likelihood that a difference is due to chance. He may also consider using more obnoxious noises if he suspects that the construction sounds were not loud enough and did not distract his participants appropriately. By making the noises louder and re-running the experiment, it may be possible to get significant results. There are many other possible answers. If you are unsure about the accuracy of your answer, feel free to ask a TA.

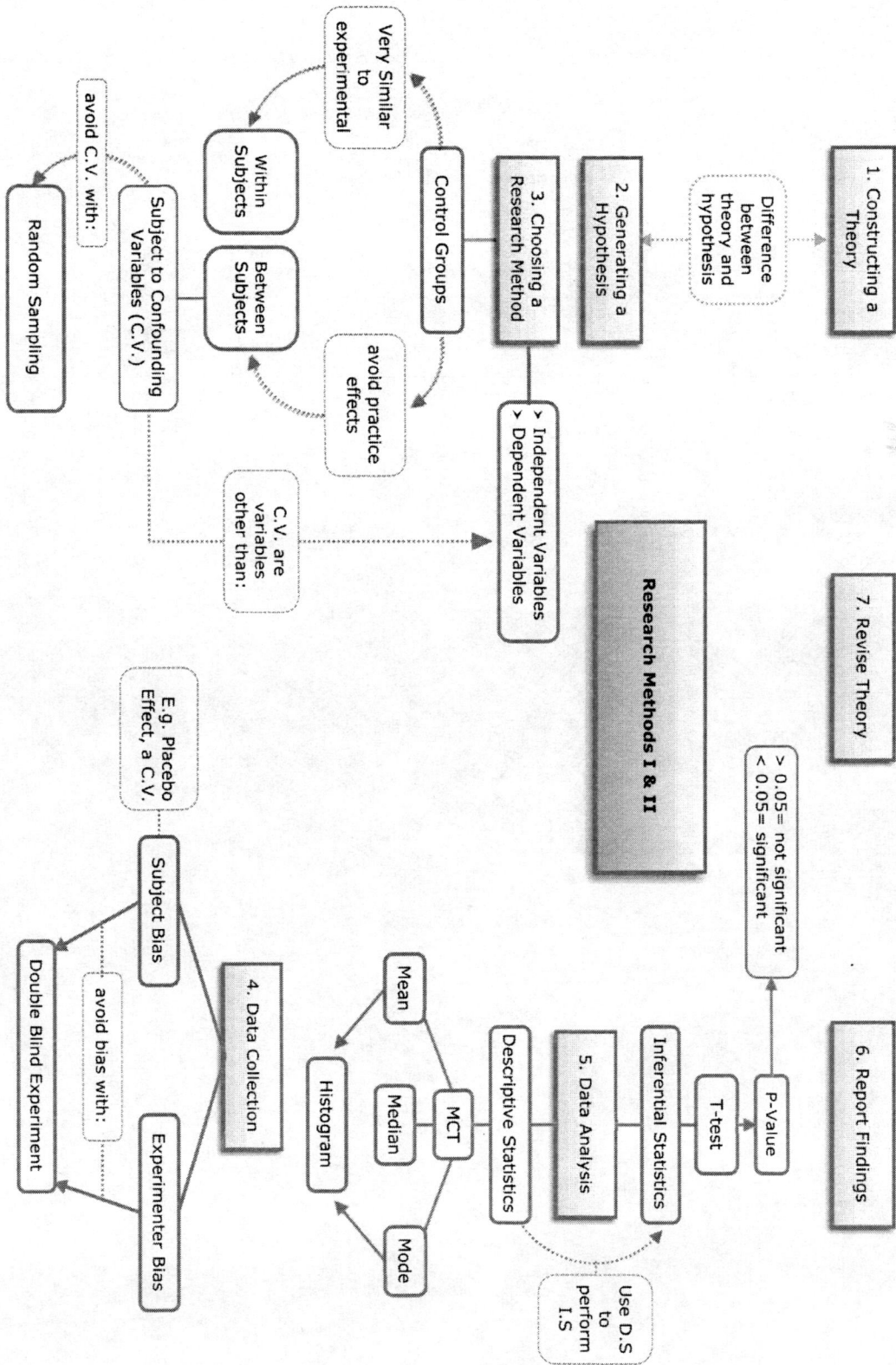

WEEK of Sept 22: CLASSICAL CONDITIONING

"We are thus confronted with two series of apparently entirely different phenomena. What now is the physiologist to do with the psychical phenomena?"
- Ivan Pavlov, 1904 Nobel Lecture address

In an episode of the television show The Office, lead character Jim describes learning in school about "this guy who trained dogs to salivate by feeding them whenever he rang a bell." For the next few weeks, Jim secretly conducts his own study using his hapless co-worker Dwight. Each time he shuts down his computer (playing the familiar Windows sound effect) Jim presents Dwight with a candy. In time, simply playing the Windows sound leads Dwight to reflexively reach for candy. The "guy who trained dogs" was a Russian physiologist named Ivan Pavlov. His experiments were known for their rigorous design and the use of objective and measurable behaviour. Pavlov's seminal work on the physiology of digestion laid the foundation for advances in theoretical and applied medicine. However, Pavlov is best known for his work on conditional reflexes conducted in the latter years of his career. In his study of the reflex regulation of the digestive glands, Pavlov paid special attention to a phenomenon that he initially called "psychic secretion." Pavlov observed that even food stimuli at a distance from the animal could elicit salivary secretions. A series of experiments led Pavlov to reject his initial subjective interpretation of psychic salivary secretion. He concluded that psychic activity was actually a reflex. More specifically, it was it a conditional reflex. This discovery was a model for how to study other so-called psychic activity through objective methods. Later experiments refined our understanding of complex interactions between an organism and its external environment. In 1904, he was awarded a Nobel Prize for Medicine/Physiology for his early work on digestive physiology. However, in his Nobel Lecture address, Pavlov chose to focus much of his speech on the conditional reflex as the first sure steps to a new line of investigation.

Pavlov, I. P. (1927). Conditioned Reflexes: An Investigation of the Physiological Activity of the Cerebral Cortex. (G. V. Anrep, Trans.). London: Oxford University Press.

Weekly Checklist:
- ☐ **Web Module to watch: Classical Conditioning**
- ☐ **Readings: Chapter 3 (Sections 1-5)**
- ☐ **AVE Quiz 3**

Upper Year Courses:
If you enjoyed the content in this week's module, consider taking the following upper year course:
- PNB 2CX3 Animal Behaviour & Evolution

Web Module 3: Classical Conditioning - Outline

Unit 1: Introduction to Learning

Two Types of Learning

Classical Conditioning

Instrumental Conditioning

Classical Conditioning and Instrumental Conditioning describe two methods of learning.

_____
_____
_____
_____
_____
_____
_____
_____
_____
_____

Unit 2: Classical Conditioning

Ivan Pavlov

Ivan Pavlov

Ivan Pavlov: _____
_____
_____
_____
_____
_____
_____
_____
_____
_____
_____
_____

Contingencies

Define contingencies: _____
_____
_____
_____
_____
_____
_____
_____
_____
_____
_____
_____
_____

Classical Conditioning

Definition

Classical Conditioning

- The learning of a contingency between a particular signal and a later event that are paired in time and/or space

Classical Conditioning can be found in many organisms and involves learning the relationship between two events.

_____
_____
_____
_____
_____
_____
_____
_____
_____

Unit 3: Terminology

The Unconditional Stimulus

Definition

Unconditioned Stimulus (US)

- Any stimulus or event
- Occurs naturally, prior to learning

Unconditional stimulus: _____
_____
_____
_____
_____
_____
_____
_____
_____
_____
_____
_____
_____
            _____

The Unconditional Response

Definition

Unconditional Response (UR)

- The response that occurs after the unconditioned stimulus
- Occurs naturally, prior to any learning

Unconditional response: _____
_____
_____
_____
_____
_____
_____
_____
_____
_____
_____
_____
_____
_____
_____

_____
_____
_____
_____
_____

The Conditional Stimulus

Definition

Conditioned Stimulus (CS)
- Paired with the unconditioned stimulus to produce a learned contingency

The Conditioned Response

Definition

Conditioned Response (CR)
- The response that occurs once the contingency between the CS and the US has been learned

Acquisition

Negatively Accelerated Increasing Function (A)

Amount of Salivation to Signal

# of trials

Conditional Stimulus: _____
_____
_____
_____
_____
_____
_____
_____
_____
_____
_____
_____

Conditioned Response: _____
_____
_____
_____
_____
_____
_____
_____
_____
_____
_____

The acquisition of a contingency typically requires many trials, but can occur quickly in certain situations.

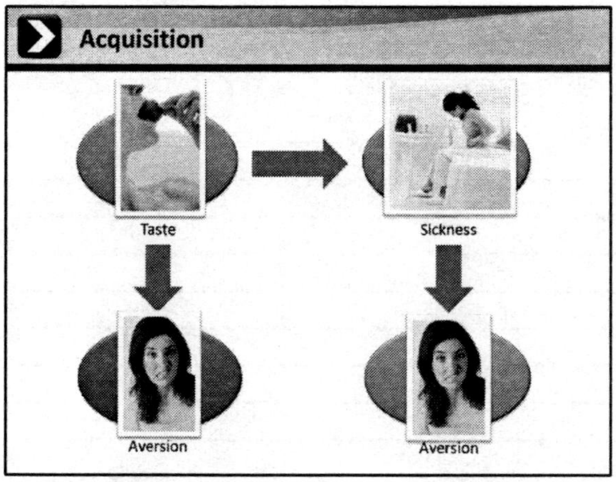

Acquisition

Taste → Sickness

Aversion   Aversion

Unit 4: Extinction

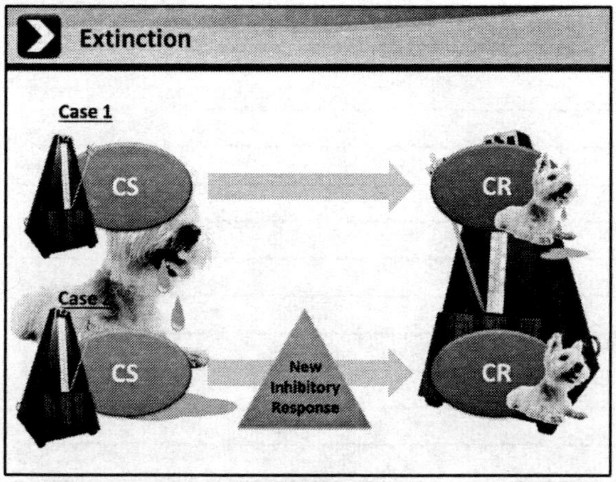

Extinction

Case 1

CS → CR

Case 2

CS → New Inhibitory Response → CR

Extinction occurs when the CS can no longer elicit a CR.

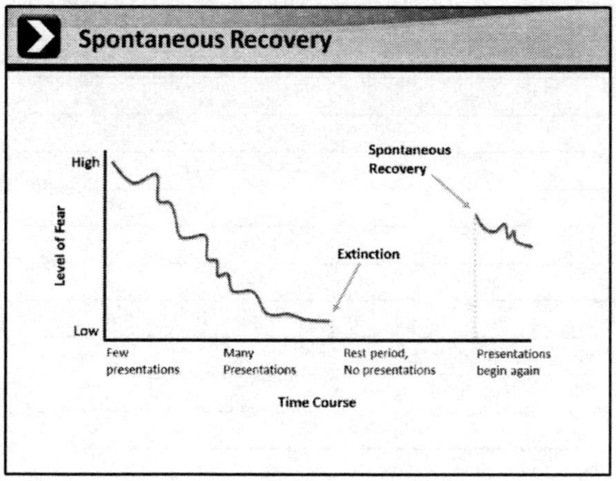

Spontaneous Recovery

Level of Fear

High

Low

Few presentations | Many Presentations | Rest period, No presentations | Presentations begin again

Spontaneous Recovery

Extinction

Time Course

Spontaneous recovery demonstrates that the original contingency is not unlearned during extinction.

Unit 5: Generalization and Discrimination

The Generalization Gradient

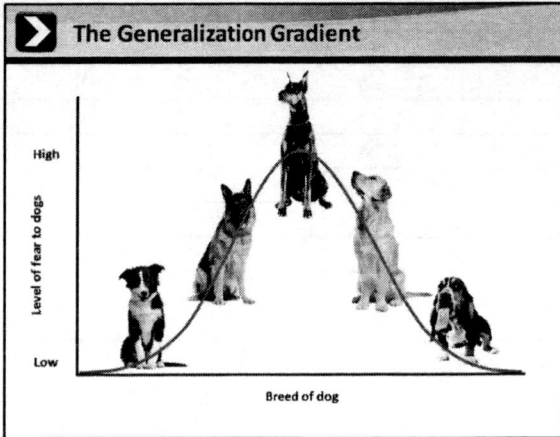

Level of fear to dogs (High / Low) vs *Breed of dog*

The CS+ and the CS-

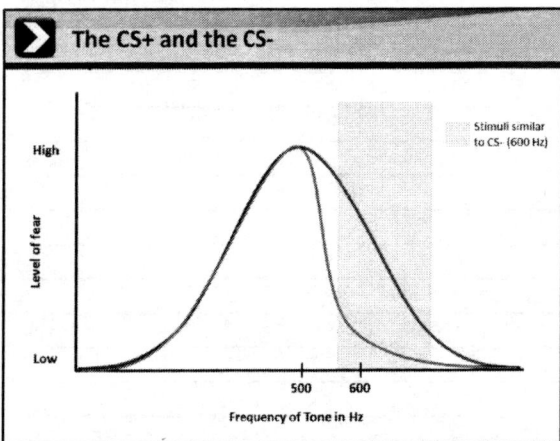

Stimuli similar to CS- (600 Hz)

Level of fear (High / Low) vs *Frequency of Tone in Hz* (500, 600)

The CS+ and the CS-

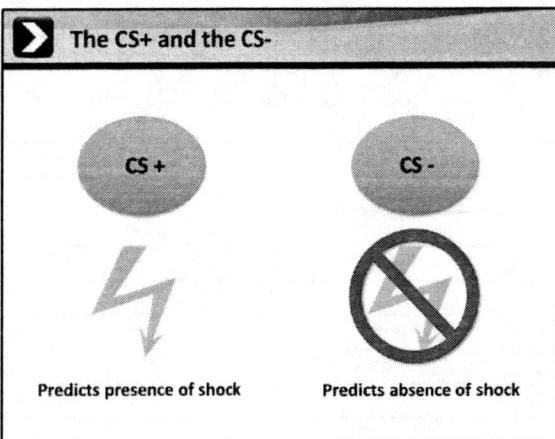

CS +

CS -

Predicts presence of shock

Predicts absence of shock

_____
_____
_____
_____
_____

Generalization occurs when the organism produces a CR to stimuli that appear similar to the original CS.

_____
_____
_____
_____
_____
_____
_____
_____
_____

Stimulus discrimination allows the organism to produce a CR to only a restricted range of stimuli.

_____
_____
_____
_____
_____
_____
_____
_____
_____
_____
_____

CS+ _____
_____
_____
_____

CS- _____
_____
_____
_____
_____
_____

_____
_____
_____
_____
_____

Conditioning and Fear

CS — Snake	US — Snake Bite
CR — Fear	UR

Phobias and Therapy

Strength of CR

Implosive Therapy & Systematic Des

Ivan Pavlov

Unit 6: Homeostasis and Compensatory Responses

The Role of Conditioning

CS — Sweet taste	US — Increase in blood sugar
CR — Insulin release	UR — Insulin release

_____
_____
_____
_____
_____
_____

Phobias and their treatment can be understood in the context of classical conditioning.

_____
_____
_____
_____
_____
_____
_____
_____
_____
_____
_____

Implosive therapy:
_____
_____
_____
_____

Systematic desensitization:
_____
_____
_____
_____
_____
_____

Homeostasis benefits greatly from classical conditioning mechanisms.

Compensatory responses:
_____
_____
_____
_____
_____
_____
_____
_____
_____
_____
_____
_____
_____
_____

Addictions

CS
Environment:

US
Drug effects:

CR
Counter-adaptations:

UR
Counter-adaptations:

Addictions and withdrawal symptoms can be understood in the context of classical conditioning.

_____
_____
_____
_____
_____
_____
_____
_____
_____

Unit 7: Conclusion

The Importance of Classical Conditioning

_____
_____
_____
_____
_____
_____
_____
_____
_____
_____
_____
_____
_____
_____
_____
_____
_____

Classical conditioning plays an important role in many aspects of our daily lives.

_____
_____
_____
_____
_____
_____
_____
_____
_____
_____
_____
_____
_____
_____
_____
_____
_____
_____
_____
_____
_____
_____
_____
_____
_____
_____
_____

Module 3: Classical Conditioning – Courseware Exercise

Imagine that you are in charge of advertising for a company that makes carbonated beverages. You have just performed a market survey and have found that you and your main competitor have approximately equal share of the market (you are both selling essentially the same amount of product). Use the principles of classical conditioning to:

1) Explain how you can mount a television ad campaign that will encourage consumers to have more positive thoughts associated with your product.

_____

_____

_____

2) Explain how you can modify your campaign to associate negative thoughts with your competitor's product.

_____

_____

_____

3) Identify the CS, US, CR, and UR and their relationship to one another in your ad campaign using the following figure from lecture.

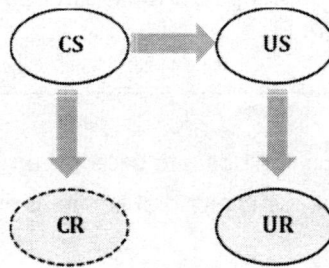

Classical conditioning can also be applied to clinical populations, particularly to those with phobias, anxiety problems, and maladaptive behaviours.

Imagine that you teach a 3rd grade class and are helping students prepare for an oral presentation. You notice, however, that a few of the students are very nervous about their presentations. One nervous student in particular has not been doing very well in your class. You fear that they may become flustered by this presentation, do poorly, and come to have a fear of all future presentations, even those in other courses.

4) What classical conditioning principle is at play in this case and how can you use stimulus discrimination to counter it? Describe how you would do so in a classroom setting.

_____

_____

_____

5) Using the techniques described in lecture, give three strategies that you could use to assure that none of your students develop a sudden fear of public speaking after this presentation.

a)_____

_____

b)_____

_____

c)_____

_____

Module 3: Classical Conditioning – Review Questions

1) In a classical conditioning framework, define:
   a) The US: _____
   b) The UR: _____
   c) The CS: _____
   d) The CR: _____

2) Describe the process of acquisition and explain what the negatively accelerating function suggests about classical conditioning.

_____
_____
_____

3) Identify the US, UR, CS, CR for the following scenarios:
   a) You eat a new food and then get sick because of the flu. However, you develop a dislike for the food and feel nauseated whenever you smell it.
      US: _____ UR: _____ CS: _____ CR: _____

   b) Tyler's date was wearing a very alluring perfume on their recent date. The date itself was quite passionate. The following day, when Tyler gets into his car, he smells the lingering scent of his date's cologne and becomes transfixed with joy.
      US: _____ UR: _____ CS: _____ CR: _____

   c) Deanna had a panic attack during a plane ride and became very afraid at the mere thought of an airplane. Twenty years later, Deanna is still afraid of airplanes even though she has never taken another flight.
      US: _____ UR: _____ CS: _____ CR: _____

4) Describe the process of generalization and discrimination that leads an individual to show a fear response to a 500Hz tone but not to a 600Hz tone.

_____
_____
_____

5) Describe the concept of extinction and spontaneous recovery and what they suggest about classical conditioning.

_____
_____
_____

6) Use the example of drinking a cola to describe the process of homeostasis and why it is important.

_____
_____
_____

7) Explain how drug addictions and withdrawal symptoms are a process of classical conditioning.

_____
_____
_____

Module 3: Classical Conditioning – Quiz Question

Which of the following correctly pairs a learning phenomenon with its example?

a. Discrimination: Xavier recalls particularly fond childhood memories whenever he smells his grandmother's pea soup, even though she cooked many other types of soup. (57.18%)
b. Habituation: Hillary is particularly afraid of her neighbour's dog after the dog bit her when they were playing once, but Hillary has no trouble interacting with any other dogs. (8.86%)
c. Generalization: Eli becomes increasingly responsive to the spontaneously scorching hot water in his shower when his roommate flushes the toilet. (9.36%)
d. Sensitization: Tanya taught her pet cockroaches to ignore a 500 Hz tone but respond to a 600 Hz tone. (24.59%)

This is an actual question from the 2011 Classical Conditioning Avenue Quiz that many students struggled with. We are going to go through this question together to identify what types of errors students made and how to avoid these mistakes in the future. To do so, we will look at each answer option individually and identify why it is correct or incorrect.

Often, exam developers will use correct information in an inaccurate way to describe or define certain concepts. In this example, various situations are described that encompass different concepts presented in the web module. It is your job to identify the correct match. Similar to questions we have walked through in the past, this question requires you to be able to understand and explain the various concepts from the web module in your own words. This will enable you to pick out the important information from each hypothetical scenario and compare it to what you've learned in the web module. Let's begin by examining the options:

a. Before reading the entire answer option we should already be defining discrimination in our own words. Once we've defined the concept we can now examine the answer option further. This option seems to be a correct example of discrimination. We will hold onto it. (**Correct!**)
b. Again, we first define habituation in our own words. Is this scenario an example of habituation? It may seem like it at first, but it is not. Habituation is the decrease in responsiveness to a stimuli due to repeated exposure with no consequences. Is this exactly what is occurring in this example? No and therefore this option is incorrect. Let's move on.
c. First we define generalization in our own words. Is this scenario an example of generalization? Generalization is the process of applying a learned contingency to a set of similar stimuli. This does not seem to be what is occurring in this scenario. Let's move on.
d. After defining sensitization as a heightened responsiveness due to increased presentation of a stimulus, we understand that this is not what is occurring in this scenario.

We are left with only **option A**. Remember, it is important to be able to understand a concept to the point where you can explain and define it in your own words. This ability will be particularly useful when applying these concepts to new situations.

Key Terms

Acquisition	Discrimination	Phobia
Addiction	Drug Tolerance	Spontaneous Recovery
Classical Conditioning	Extinction	Systematic Desensitization
Compensatory Responses	Generalization	Taste Aversion
Conditioned Response (CR)	Generalization Gradient	Unconditional Response (UR)
Conditioned Stimulus (CS)	Homeostasis	Unconditional Stimulus (US)
Contingency	Implosive Therapy	Withdrawal
CS+	Inhibition	
CS-	Overdose	

Module 3: Classical Conditioning – Bottleneck Concepts

Drug Overdose vs. Withdrawal
CS+ vs. CS-
Homeostasis
Spontaneous Recovery

Drug Overdose and Withdrawal

To understand drug overdose and withdrawal from a classical conditioning perspective, you must first understand homeostasis. Briefly, homeostasis is the process of keeping the internal environment of the body constant. When someone takes a drug, it changes their body in some way. For example, the drug might increase their heart rate. The body, wanting to maintain a constant internal environment, will in turn decrease the individual's heart rate to counteract the effects of the drug.

Often, people take drugs in a similar environment (e.g. for smokers: outside of their office building, using a certain brand of cigarettes, during their break at work etc.). Over time, the environment becomes associated with consuming the drug and experiencing its effects. This means that being placed in the drug environment alone can elicit a conditioned response of counter-adaptations (processes that counteract the drug effects). This is the basis of both drug overdose and drug withdrawal.

When taking a drug in the same environment as usual, counter-adaptations occur before the drug is ingested. These measures prepare the body for the effects of the drug, leading a certain dose of drug to elicit a smaller effect. However, if the person takes the drug in an unfamiliar environment, counter-adaptations will not occur. With a lack of countermeasures preparing the body, the same dose of drugs may lead to overdose when taken in an unfamiliar environment. In terms of drug withdrawal, being in the drug environment or seeing the apparatus needed to ingest a drug (e.g. a syringe) will elicit counter-adaptations. If no drug is taken, these counter-adaptations will cause the body to be abnormal (e.g. if a drug makes you ecstatic, the counter-adaptation might make you depressed). These counter-adaptations often make a person feel uncomfortable and may explain some of the withdrawal symptoms people experience.

Students often get confused about how these concepts work. A good way to think through these problems is to draw the CS, CR, US, and UR diagram (see the exercises below). Then, you can ask yourself: What happens during withdrawal when you are in an environment associated with drugs, but did not take the drug? You would still experience the CR, but without drugs to counteract the effects, you would experience withdrawal. On the other hand, what do you suppose happens during a drug overdose where you have the drug effects but not the CS? There would be a UR but no CR to help the body adapt to the drug effects!

Test your Understanding

1) Draw the US, UR, CS, CR diagram for the contingency between the environment and drug effects. Use this diagram to explain why drug withdrawal and drug overdose occur.

Answer:

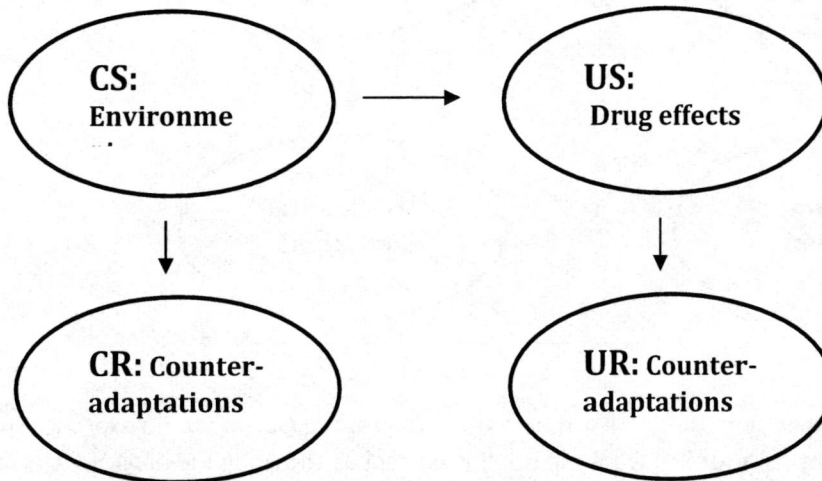

In drug withdrawal, the CS is presented without the US. The CS produces the CR of counter-adaptations which, in the absence of drug effects, causes the body to become abnormal. This may cause discomfort and result in withdrawal.

In drug overdose, a high amount of US is presented without the CS. Since the CS is unavailable, it cannot elicit the CR and the body is unprepared for a high dose of drugs. This situation may cause an unusually elevated drug effect for which the UR alone cannot compensate, leading to overdose.

2) Gordon works out daily at the No Pain No Gain gym. After a while at the gym, he starts being able to lift 200lbs without even having to warm-up. According to classical conditioning, which of the following statements is incorrect?

a) When going to a different gym, Gordon will likely not be able to bench-press as much weight without first warming up.
b) The No Pain No Gain gym has become associated with doing exercise. Therefore, being present at the gym leads to Gordon's body preparing itself for exercise.
c) The No Pain No Gain gym environment has become the CS for counter-adaptations, the US.
d) If the No Pain No Gain gym undergoes a complete renovation, Gordon may not be able to lift as much weight without warming up.

The answer is C. Begin by drawing the classical conditioning diagram:

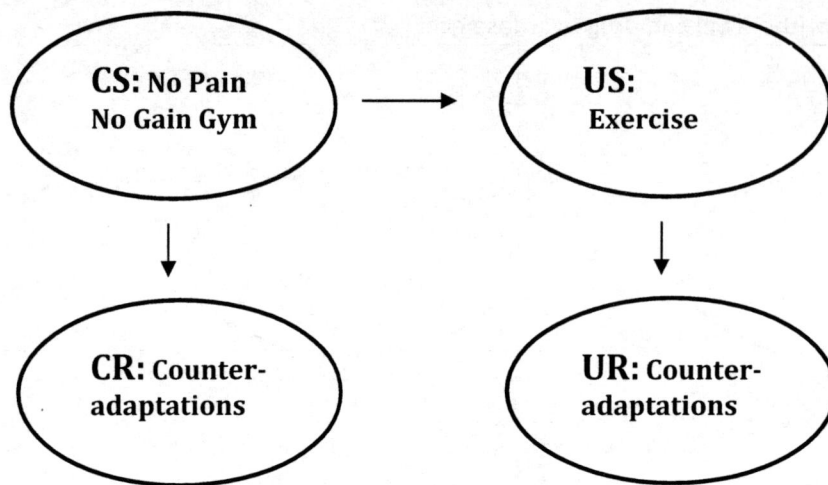

A is correct because in any other gym, there will not be any CR to prepare Gordon to do exercise. Thus, he would likely be able to lift less weight without first warming up. B is correct as shown in the diagram. C is incorrect because exercise is the US which leads to the UR, counter-adaptations. D is correct because if the gym is completely renovated, the environment will no longer be the same. Losing the CS would no longer prepare Gordon's body for exercise.

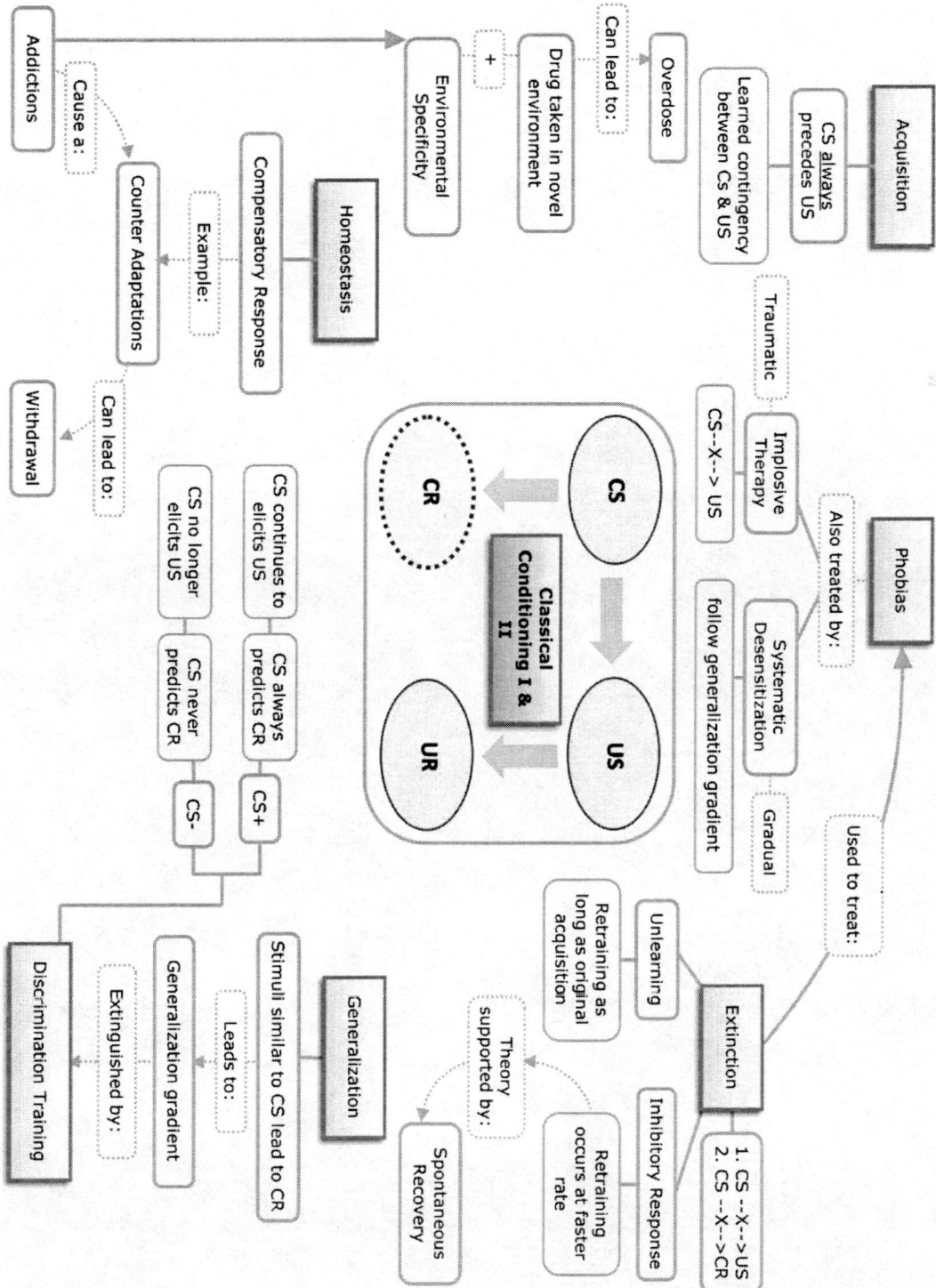

WEEK of Sept 29: INSTRUMENTAL CONDITIONING

"Give me a dozen healthy infants, well-formed, and my own specified world to bring them up in and I'll guarantee to take any one at random and train him to become any type of specialist I might select— doctor, lawyer, artist, merchant-chief and, yes, even beggar-man and thief, regardless of his talents, penchants, tendencies, abilities, vocations, and race of his ancestors. "
-John B. Watson, author of "Psychology as the Behaviourist Views It."

For a number of years, there was an unusual display in the lobby of the psychology building at Harvard University. Remarkably, two pigeons would hit a ping-pong ball back and forth to each other in a spirited match. Visitors were amazed at how such a complex skill could be taught to birds. Although impressive, training was actually straightforward and less complex than people realized. B.F. Skinner first trained each pigeon individually. The desired behaviour of hitting a moving ping-pong ball was broken down into particular bits of behaviour and reinforced (rewarded) with food. The bird might first be reinforced for just approaching the ball, then only for touching the ball, later only for pushing it, and finally only for hitting a moving ball. All that was left was to place the two trained pigeons at opposite ends of a table and start keeping score. By a similar process, Skinner taught pigeons to walk in figure-eights, dance with each other, and even operate a prototype missile guidance system (Skinner lamented that "no one would take us seriously.") The core principles that actions are determined by the environment and that behaviour is shaped and maintained by its consequences are used in many applied settings including cognitive behaviour therapy, game design, and educational psychology. However, Skinner went a step further and believed that these foundational principles could introduce positive change to our society as a whole. As summarized by one interviewer, "In the Skinnerian world, man will refrain from polluting, from overpopulating, from rioting, and from making war, not because he knows that the results will be dangerous, but because he has been conditioned to want what serves group interests."

Watson, J.B. (1913). Psychology as a behaviourist views it. Psychological Review, 20, 158-177. Skinner, B.F. (1950). Are theories of learning necessary? Psychological Review, 57, 193-216.

Weekly Checklist:
- ☐ **Web Module to watch: Instrumental Conditioning**
- ☐ **Readings: Chapter 3 (Sections 6-8)**
- ☐ **AVE Quiz 4**

Upper Year Courses:
If you enjoyed the content in this week's module, consider taking the following upper year course:
PSYCH 2H03 Human Learning and Cognition

Module 4: Instrumental Conditioning – Outline

Unit 1: Introduction

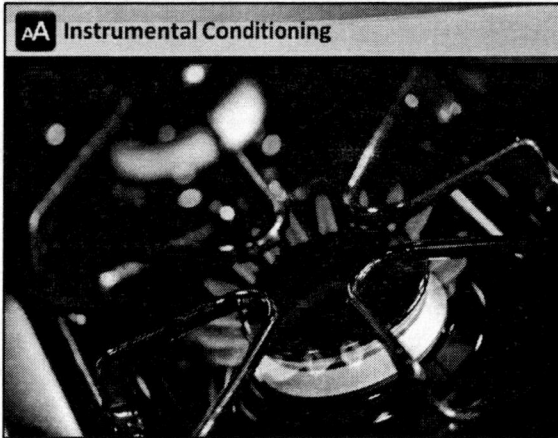

Instrumental conditioning: _____
_____
_____
_____
_____
_____
_____
_____
_____
_____
_____
_____

Unit 2: Instrumental Conditioning

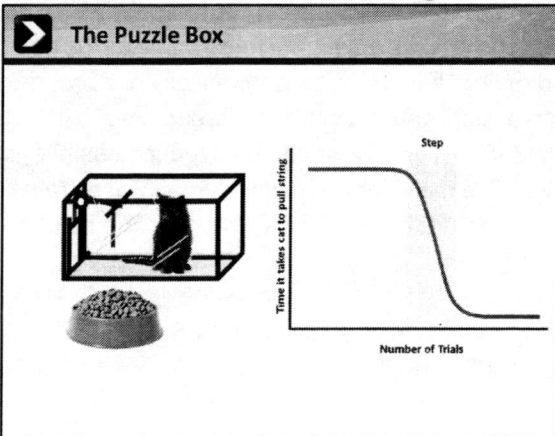

Early research on instrumental conditioning was often performed on non-human animals.

Thorndike: _____
_____
_____
_____
_____
_____
_____
_____
_____
_____

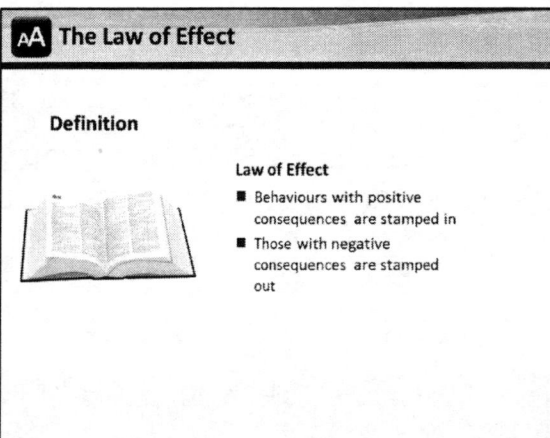

Stamping in: _____
_____
_____
_____

Stamping out: _____
_____
_____
_____
_____
_____

Unit 3: Types of Instrumental Conditioning

Four Consequences

Reward Training

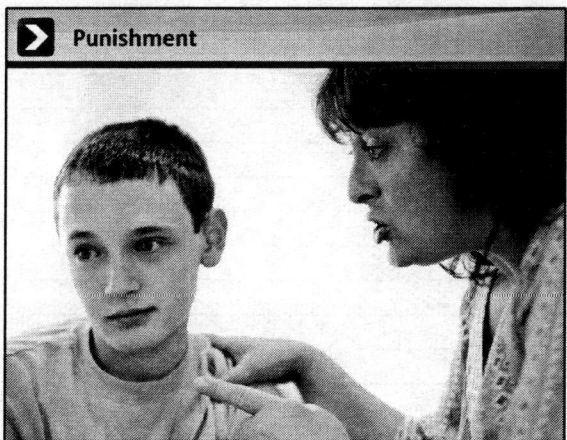

Punishment

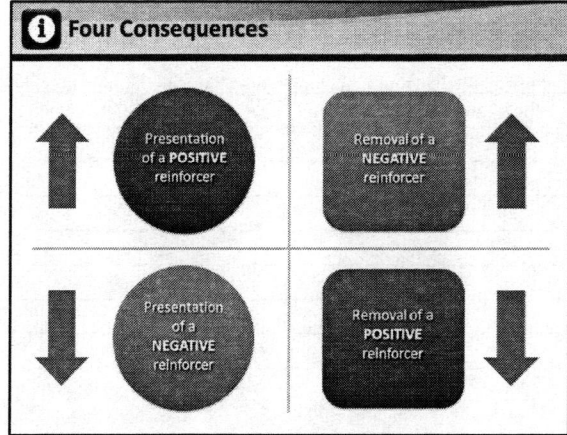

_____
_____
_____
_____
_____
_____

Behaviour can be influenced by four types of consequences.

_____
_____
_____
_____
_____
_____
_____
_____
_____
_____
_____

Reward training:

_____
_____
_____
_____
_____
_____
_____
_____
_____
_____

Punishment:

_____
_____
_____
_____
_____
_____
_____
_____
_____
_____
_____
_____
_____

Omission Training

Removal of a **POSITIVE** reinforcer

MOOOMMM!!

You stiiiiink! Nah nah nah nah naah!

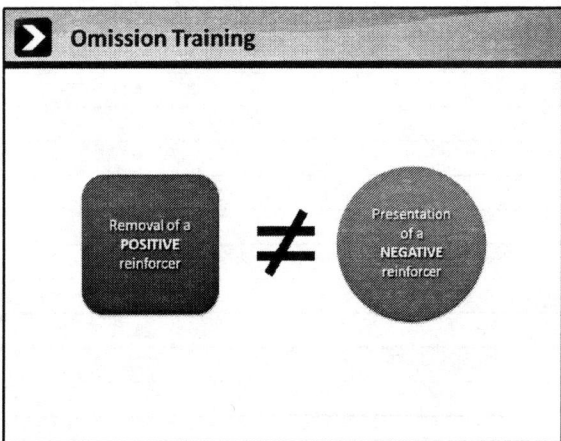

Omission Training

Removal of a **POSITIVE** reinforcer ≠ Presentation of a **NEGATIVE** reinforcer

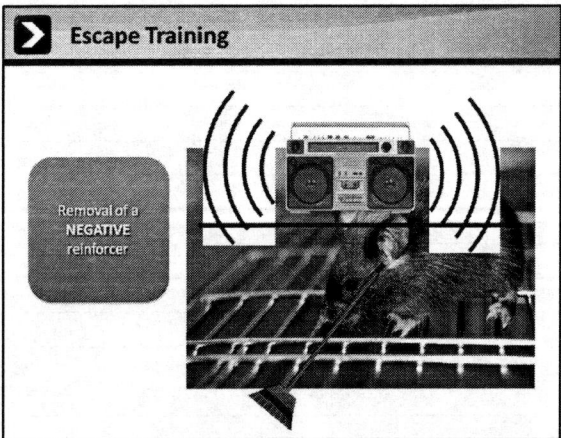

Escape Training

Removal of a **NEGATIVE** reinforcer

Omission training:
_____
_____
_____
_____
_____
_____
_____
_____
_____
_____
_____

_____
_____
_____
_____
_____
_____
_____
_____
_____
_____
_____

Escape training:
_____
_____
_____
_____
_____
_____
_____
_____
_____
_____
_____
_____
_____
_____
_____
_____
_____

_____
_____
_____
_____
_____
_____

Unit 4: Acquisition and Shaping

Graphing Responses

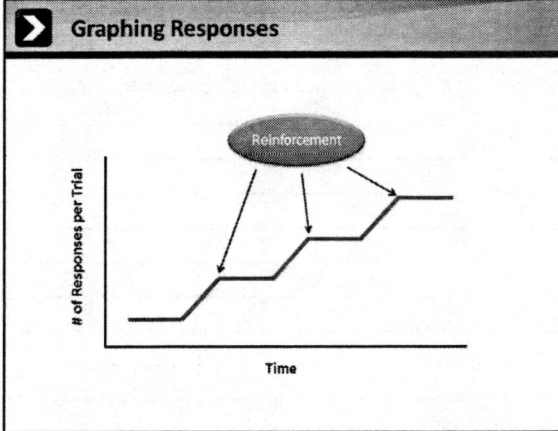

Responses to each type of training can be graphed to visualize the acquisition of a contingency.

_____
_____
_____
_____
_____
_____
_____
_____
_____
_____
_____

Autoshaping

Organisms can learn the contingency between a behaviour and its consequence without explicit training.

Autoshaping: _____
_____
_____
_____
_____
_____
_____
_____
_____
_____
_____

Shaping

_____
_____
_____
_____
_____
_____
_____

Some behaviours require a more systematic form of training to enable the acquisition of a contingency.

Shaping by successive approximation: _____
_____
_____
_____
_____
_____
_____
_____
_____
_____
_____
_____
_____
_____
_____

Unit 5: Generalization and Discrimination

The Discriminative Stimulus

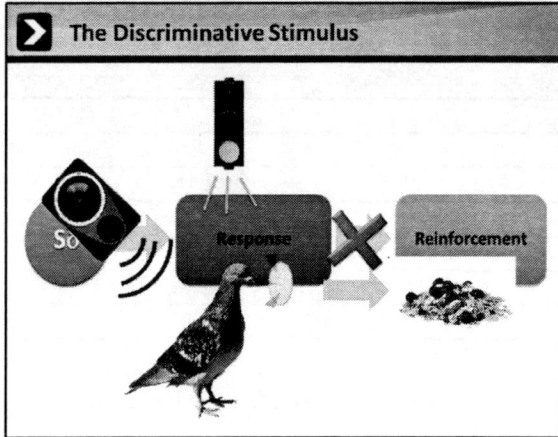

Generalization

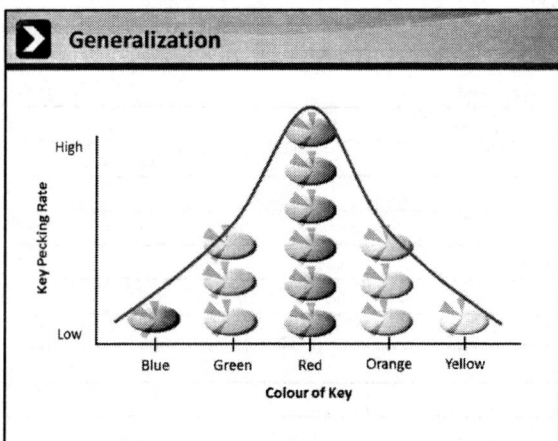

SD and S-Delta

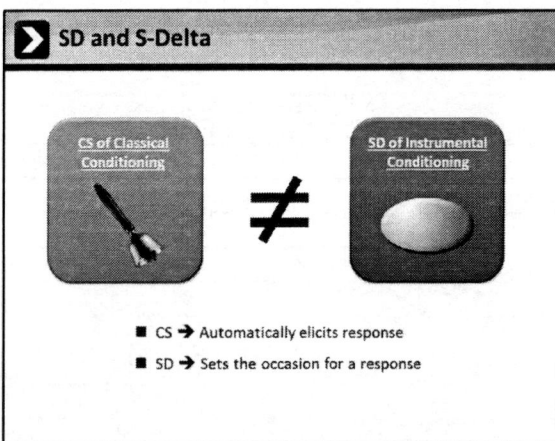

- CS → Automatically elicits response
- SD → Sets the occasion for a response

_____
_____
_____
_____
_____
_____
_____

The discriminative stimulus indicates the validity of a contingency.

SD: _____
_____
_____
_____

S-delta: _____
_____
_____
_____
_____

Stimuli similar to the SD can indicate the validity of a contingency to a certain degree.

_____
_____
_____
_____
_____
_____
_____
_____
_____
_____
_____

The ability to elicit a response involuntarily marks an important distinction between the CS and the SD.

_____
_____
_____
_____
_____
_____
_____
_____
_____
_____
_____
_____
_____
_____

Unit 6: Schedules of Reinforcement

Four Schedules

Four basic schedules of reinforcement:

Fixed Ratio (FR-#)	Variable Ratio (VR-#)
Fixed Interval (FI-#)	Variable Interval (VI-#)

Fixed Ratio

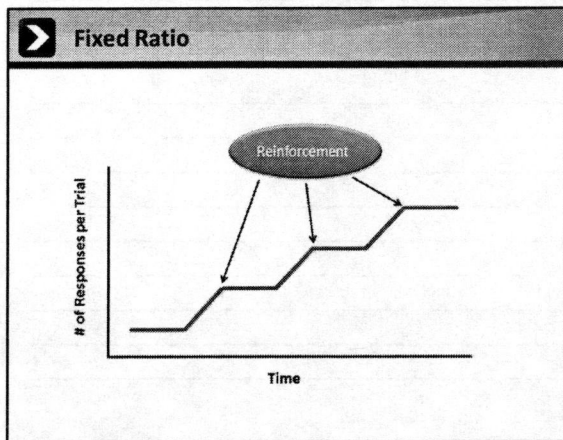

Variable Ratio

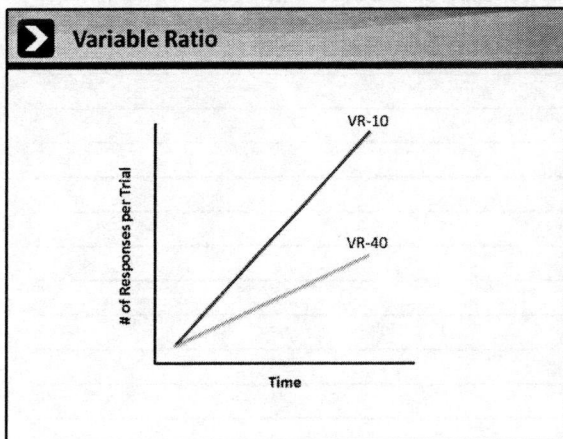

Four basic schedules guide the delivery of reinforcement.

_____
_____
_____
_____
_____
_____
_____
_____
_____
_____

Fixed ratio: _____
_____
_____
_____
_____
_____
_____
_____
_____
_____
_____
_____

Variable ratio: _____
_____
_____
_____
_____
_____
_____
_____
_____
_____
_____
_____
_____
_____
_____
_____

Fixed Interval

Fixed interval:

_____
_____
_____
_____
_____
_____
_____
_____
_____
_____

Variable Interval

Variable interval:

_____
_____
_____
_____
_____
_____
_____
_____
_____
_____
_____

The schedule of reinforcement must be carefully chosen to ensure long-lasting learning.

Extinction and Schedules

_____
_____
_____
_____
_____
_____
_____
_____
_____
_____
_____
_____
_____
_____
_____
_____
_____
_____
_____

Module 4: Instrumental Conditioning – Courseware Exercise

1) Name and describe each of the four types of instrumental conditioning and provide an example of how a parent can use each technique to train her child, Billy, to pick his jacket off the floor.

a) Name: _____
   Description: _____
   Example: _____

b) Name: _____
   Description: _____
   Example: _____

c) Name: _____
   Description: _____
   Example: _____

d) Name: _____
   Description: _____
   Example: _____

e) Despite using these techniques, the parent is having no luck with the child at all! Based on the information provided in lecture, what might the child's parent be doing wrong? How could the parent fix this problem?

   _____
   _____
   _____
   _____

2) Carrie, a first year psychology student, is backpacking across Europe during the summer. She is running low on funds and decides to work at an organic strawberry farm as a fruit picker. She is paid 15 Euros for every 3 buckets of strawberries that she picks. Based on this information, respond to the following questions:

a) What is her schedule of reinforcement?

   _____

b) If Carrie begins working for another farm that pays her 15 Euros for every 2 buckets of strawberries, what is her new schedule of reinforcement?

   _____

c) How will Carrie's rate of work change with this pay increase?

   _____

d) Carrie's employers are considering the possibility of paying their workers hourly. If they decide to pay their employees 15 Euros per hour, what would her new schedule of reinforcement be?

   _____

Module 4: Instrumental Conditioning – Review Questions

1) Compare and contrast classical and instrumental conditioning.

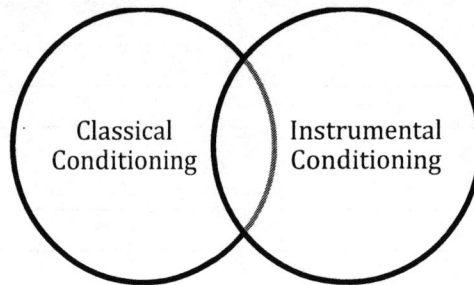

2) Which of the following definitions best describes the law of effect in instrumental conditioning?
   a) When contingencies are fully established between behaviours that lead to positive or negative outcomes, these behaviours are stamped in.
   b) Behaviours that do not lead to any outcomes are stamped in and behaviours that lead to negative outcomes are stamped out.
   c) Behaviours that lead to positive outcomes are stamped in and behaviours that lead to negative outcomes are stamped out.
   d) Behaviours that lead to rewards are stamped in and behaviours that that lead to punishment are stamped out.

3) Sally, your friend, always tries to talk to you about her weekend while the IntroPsych TA is reviewing the course concepts. This is really distracting and you want her to stop. Being the expert in instrumental conditioning that you now are, you offer her a piece candy when she stops talking. What kinds of instrumental conditioning are now associated with candies and talking in IntroPsych tutorials?
   a) For **Sally**, this is a form of **punishment**; for **you**, this is a form of **escape training**.
   b) For **Sally**, this is a form of **reward training**; for **you**, this is a form of **escape training**.
   c) For **Sally**, this is a form of **omission training**; for **you**, this is a form of **reward training**.
   d) For **Sally**, this is a form of **escape training**; for **you**, this is a form of **punishment**.

4) How does SD differ from S-delta?
   a) SD is a stimulus that will lead to positive reinforcement; S-delta is a stimulus that will lead to negative reinforcement.
   b) SD is a stimulus that signals that a behaviour will be followed by reinforcement; S-delta is a stimulus that signals that a behaviour will not be followed by reinforcement.
   c) SD automatically elicits a response; S-delta sets the occasion for a response.
   d) SD is a stimulus that is associated with generalization; S-delta is a stimulus that is associated with discrimination.

5) Graph the typical response patterns of each of the schedules.

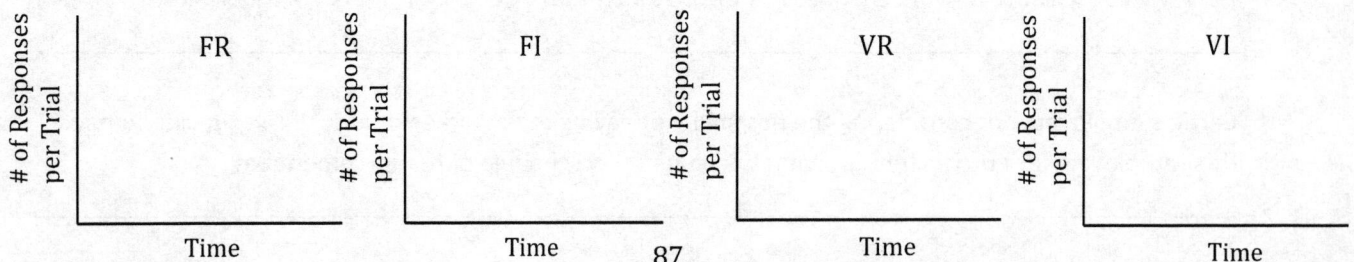

Module 4: Instrumental Conditioning – Quiz Question

Which of the following is a **correct** comparison of classical and instrumental conditioning?

a. The S-Delta cues a response to negative reinforcement, while the CS- signals the absence of a stimulus. (14.25%)
b. The S-Delta cues the absence of a response, while the CS- signals a negative response. (7.06%)
c. The SD indicates a valid contingency, while the CS+ elicits a conditioned response. (62.04%)
d. The SD indicates a positive response, while the CS+ cues a conditioned response. (16.65%)

This is an actual question from the 2011 Instrumental Conditioning Avenue Quiz that many students struggled with. We are going to go through this question together to identify what sort of errors students were making and how to avoid them in the future. To do so, we will look at each answer option individually and identify why it is correct or incorrect.

Similar to what we have done in the past, it is important that we are able to explain and describe the concepts from the web module in our own words. In addition, be sure to read every answer option carefully. Reading an answer option with a critical and sceptical eye will allow you to better identify the inconsistencies in an option. Keeping this in mind, let's examine the answer options:

a. In this case, we should be able to describe a CS+, CS, discriminative stimulus and S-Delta. Furthermore, we are told the S-Delta cues a response. Have we ever talked about cueing a response in either type of conditioning? No. In fact, the S-Delta is a signal that a particular contingency is not valid. This option is incorrect. Let's move on.
b. Again, does the description of the S-Delta match the description we have given in our words? No. What about the CS-? Well, the CS- signals the absence of a stimulus (US). Is this how the CS- is described in this question? No. It is described as a predictor of a negative response. It may sound right but we can identify it as being incorrect when we compare it to our description of the CS-. Let's move onto option C.
c. Does the description of the S-Delta match ours? Yes! What about the CS+? Well, sort of. While not explicitly mentioning that the CS+ predicts the presence of the US, it mentions that it will elicit a UR. We know that if a CS predicts the presence of a US, then once a CS-US contingency is learned, the CS will elicit a CR. This option seems to be correct so we will hold onto it. (**Correct!**)
d. This option states that the SD indicates a positive response. What does that even mean? The SD makes you feel good? Makes you perform a good deed? This is a nonsense statement designed to sound like a good option. Remember, it is important to be critical of each answer option. The correct option can always be supported by information from the web module without having to "stretch" the information to fit the option. Therefore, this is an incorrect alternative.

This leaves us with **option C,** the correct answer in this question. Often, answer options are designed to sound like good answers. If we take the time to think critically and apply the information we have learned from the web module, it is easy to pick out these answer options as incorrect.

Key Terms

Acquisition	Four Consequences	Puzzle Box
Autoshaping	Interval Schedule of Reinforcement	Ratio Schedule of Reinforcement
Continuous Reinforcement	Instrumental Conditioning	Reward Training
Discriminative Stimulus (SD)	Law of Effect	S-delta
Escape Training	Omission Training	Shaping
Fixed Interval	Partial Reinforcement	Variable Interval
Fixed Ratio	Punishment	Variable Ratio

Module 4: Instrumental Conditioning – Bottleneck Concepts

Classical Conditioning vs. Instrumental Conditioning
Omission vs. Escape Training
Schedules of Reinforcement
SD vs. S-delta

Classical Conditioning vs. Instrumental Conditioning

Both classical and instrumental conditioning are based on learning a contingency — an association or relationship between two things. Classical conditioning involves learning a contingency between two events or stimuli. Instrumental conditioning involves learning a contingency between a behaviour and its consequence. For classical conditioning, the two stimuli can be unrelated events; however, they must occur close to each other in both time and space. For instrumental conditioning, the behaviour leads to a certain consequence, meaning that the two stimuli are related in some way.

Students often have trouble identifying whether a situation is an example of classical or instrumental conditioning. The easiest way to differentiate between them is to ask whether the conditioning results in a voluntary or an involuntary response. In classical conditioning, the response (CR) is always involuntary. If you get bitten by a dog and, as a result, associate pain with dogs, you will experience fear every time you see a dog whether you want to or not. In Instrumental conditioning, however, the subject's response is always voluntary. You might know that not studying will lead to poor marks on a test, but that does not automatically mean you will study hard.

Test your Understanding

Read the following scenarios and determine if they are examples of classical or instrumental conditioning. If you classify a scenario as classical conditioning, determine the US, CS, UR, and CR. If the situation fits the description of instrumental conditioning, determine the behaviour and its consequence.

a) Tom is a curious boy who likes to explore his surroundings. One day he notices an electrical socket and wonders what it does. He decides to stick a paperclip into the hole and receives a painful shock. Ever since that day, Tom has never stuck anything into an electrical socket again.

Answer: This is an example of instrumental conditioning. Tom has learned the contingency between a behaviour (sticking things into an electrical socket) and a consequence (electrical shock/pain). He knows that sticking things into the socket leads to a negative consequence and voluntarily avoids doing the action again.

b) Francine's favourite fruit is watermelon and she eats it throughout the day. One day, she became very ill from food poisoning and felt terribly nauseous and sick. Once she felt better, Francine wanted to eat watermelons again. However, this time she felt nauseous as soon as she saw the fruit.

Answer: This is an example of classical conditioning. Francine has learned the contingency between two stimuli, watermelon and food poisoning. The feeling of nausea is involuntary.

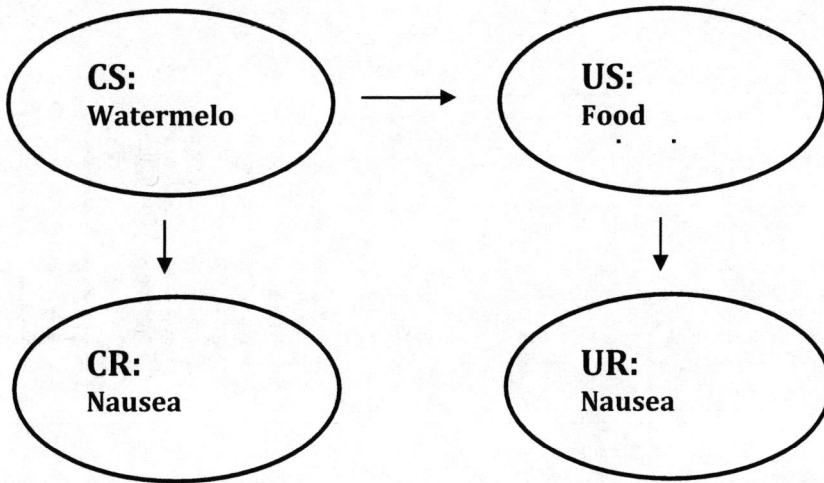

2) Come up with an example of instrumental and classical conditioning based on your everyday life. If you are unsure about their accuracy, feel free to ask your TA for their opinion!

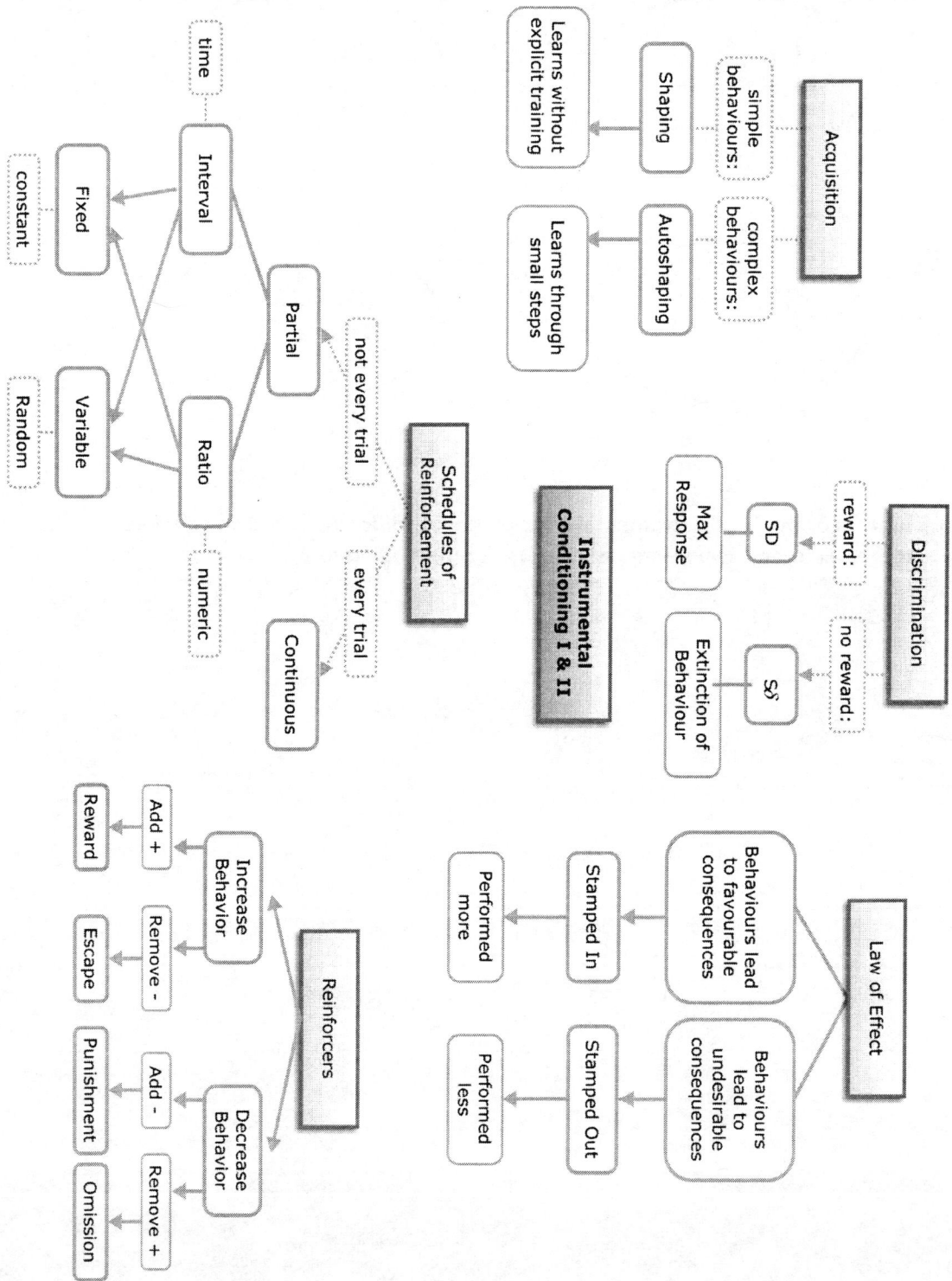

WEEK of Oct 6: PROBLEM SOLVING & INTELLIGENCE

"If I had an hour to solve a problem and my life depended on the solution, I would spend the first 55 minutes determining the proper question to ask. For once I know the proper question, I could solve the problem in less than five minutes."
- Albert Einstein

In the TV game show, *Are You Smarter than a 5th Grader?,* adult contestants competed to win a million dollar prize. In order to win, players were required to answer ten questions correctly. The questions were taken from elementary school textbooks and ranged in difficulty from grade one to grade five. Although the premise of this game seems simple, there were only two winners in the history of the show. These surprising results lead us to ask, "are adults *not* smarter than 5th graders?".

A dramatic and potentially controversial finding in intelligence research would suggest that no, adults are not smarter than current fifth graders. This finding, termed the Flynn Effect, states that the mean score for intelligence testing in the population has been steadily increasing since it was first measured in 1932. It proposes that IQ test scores are steadily on the rise, with each generation becoming more intelligent than the previous one. Despite the many ideas set forth to explain the Flynn effect, no one is actually sure if and why it occurs. Is it the result of better nutrition and better quality schooling? Or perhaps is it due to our ever-increasing access to information? Although its cause has yet to be determined, the Flynn effect is significant as it raises important questions in our minds. Specifically, it makes us wonder what contributes to this abstract and often confusing concept we call "intelligence".

Flynn, J. R. (2006). O efeito Flynn: Repensando a inteligencia e seus efeitos [The Flynn Effect: Rethinking intelligence and what affects it]. In C. Flores-Mendoza & R. Colom (Eds.), Introducao a Psicologia das Diferencas Individuais [*Introduction to the psychology of individual differences*] (pp. 387–411). Porto Alegre, Brazil: ArtMed.

Weekly Checklist:
- ☐ **Web Module to watch: Problem Solving & Intelligence**
- ☐ **Readings: Journal Article**
- ☐ **AVE Quiz 5**

Upper Year Courses:
If you enjoyed the content in this week's module, consider taking the following upper year courses:
- PSYCH 2H03 Human Learning and Cognition
- PSYCH 3I13 Cognitive Development

Module 5: Problem Solving & Intelligence – Outline

Unit 1: Introduction

Operational Definition of Intelligence

Definition

Intelligence
- The cognitive ability of an individual to learn from experience, reason well, remember important information, and cope with the demands of daily living

Intelligence: _____
_____
_____
_____
_____
_____
_____
_____
_____
_____
_____

Unit 2: Problem Solving

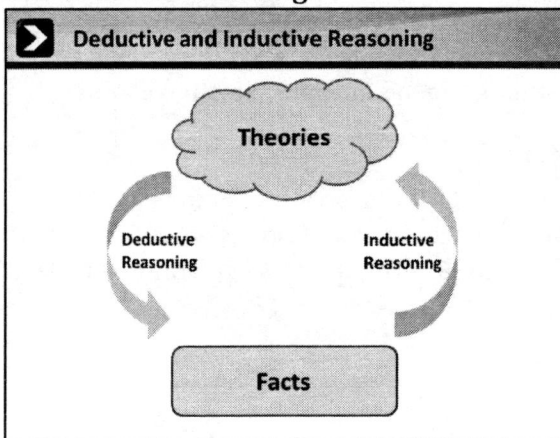

Deductive and Inductive Reasoning

Theories

Deductive Reasoning   Inductive Reasoning

Facts

We use two main types of reasoning to explain the world around us.

Deductive: _____
_____
_____
_____

Inductive: _____
_____
_____
_____
_____
_____

Insight Problems

Insight problems are a special type of puzzles that engage our problem solving skills.

Functional fixedness: _____
_____
_____
_____
_____
_____
_____
_____
_____
_____

Unit 3: A History of Intelligence Testing

The Qualities of a Test

Reliability

A reliable test produces the same result if one person takes it multiple times

The Qualities of a Test

Validity

A valid test measures only the trait it is supposed to be measuring

SAT
MCAT
GRE

Francis Galton

Increased reaction time

Higher intelligence?

Francis Galton

_____
_____
_____
_____
_____
_____

An effective test of intelligence must be both reliable and valid.

Reliability: _____
_____
_____
_____
_____
_____
_____
_____
_____
_____
_____

Validity: _____
_____
_____
_____
_____
_____
_____
_____
_____
_____
_____
_____

Historically, intelligence has been tested in several different ways and each method has employed a variety of assumptions.

Francis Galton: _____
_____
_____
_____
_____
_____
_____
_____
_____
_____
_____
_____
_____
_____
_____

Charles Spearman & 'G'

- Intelligence tests
- Vocabulary
- Math
- Special Abilities
- ...

→ G

Multiple Intelligences

- Linguistic – Verbal
- Mathematical – Logical
- Rhythmic – Musical
- Spatial – Visual
- Kinesthetic – Bodily
- Interpersonal
- Intrapersonal
- Naturalistic

Check out +Docs to try a Multiple Intelligence test based on Gardner's Theory of Intelligence

Unit 4: Human Intelligence

> **The Weschler Scales**

> **Genetic and Environmental Contributions**

Identical
(100% identical genes)

Fraternal
(50% identical genes)

McGue et al. (2003) Correlation between IQs of twins:

- Identical = 0.80
- Fraternal = 0.60

> **Genetic and Environmental Contributions**

IQ correlation = 0.73

_____
_____
_____
_____
_____
_____

IQ tests remain popular today and allow scientists to research intelligence.

Characteristics of modern-day IQ tests: _____

_____
_____
_____
_____
_____
_____
_____
_____
_____
_____

Methods used to determine the influence of genes and environment on intelligence: _____

_____
_____
_____
_____
_____

Findings from twin studies: _____

_____
_____
_____
_____

Implications of findings: _____

_____
_____
_____
_____
_____
_____
_____
_____
_____
_____
_____
_____
_____
_____
_____

The Flynn Effect

Definition

The Flynn Effect
- Raw IQ scores have been on the rise since 1932

The Flynn Effect: _____
_____
_____
_____
_____

Potential reasons for the Flynn Effect: _____
_____
_____
_____
_____
_____

Unit 5: Piaget and Intelligence Development

Jean Piaget

Definition

Schema
- A mental framework for interpreting the world around us

Assimilation
- Incorporation new information into existing schemas

Accommodation
- Modifying existing schemas to fit incompatible information

Jean Piaget suggested theories of intelligence development that remain influential to this day.

Schema: _____
_____
_____

Assimilation: _____
_____
_____

Accommodation: _____
_____
_____

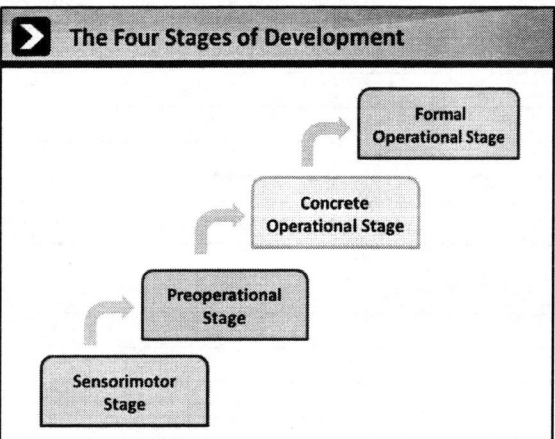

The Four Stages of Development

Formal Operational Stage

Concrete Operational Stage

Preoperational Stage

Sensorimotor Stage

Characteristics of the four stages of development: _____
_____
_____
_____
_____
_____
_____
_____
_____
_____
_____
_____

_____
_____
_____
_____
_____
_____

_____
_____
_____
_____
_____
_____

The Sensorimotor Stage

Object Permanence
- Realization that objects continue to exist when no longer visible

0-2 years 2-7 years 7-11 years 11+ years

The Preoperational Stage

Object Permanence ✓

Egocentrism

Seriation

Reversible relationships

Conservation

0-2 years **2-7 years** 7-11 years 11+ years

The Preoperational Stage

Conservation

Check out +Media for a demonstration of the Conservation Task

= → <

0-2 years **2-7 years** 7-11 years 11+ years

_____
_____
_____
_____
_____
_____
_____
_____

The Sensorimotor Stage

Age: _____
_____

Milestones: _____
_____
_____
_____
_____

Additional information: _____
_____
_____

The Preoperational Stage

Age: _____
_____

Egocentrism: _____
_____
_____

Seriation: _____
_____
_____
_____

Reversible relationships: _____
_____
_____
_____

Conservation: _____
_____
_____
_____
_____
_____
_____
_____
_____
_____

The Concrete Operational Stage

Age: _____
_____

Characteristics: _____
_____
_____
_____
_____
_____
_____
_____

_____

Formal Operational Stage

Age: _____
_____

Characteristics: _____
_____
_____
_____
_____
_____
_____
_____

_____
_____
_____
_____
_____
_____
_____
_____

Criticisms of Piaget's Theories

Decalage: _____
_____
_____
_____

Children's language abilities: _____
_____
_____
_____
_____
_____
_____
_____
_____
_____
_____
_____
_____
_____
_____

Unit 6: Biases and Heuristics

Confirmation Bias

The Availability Heuristic

The Representativeness Heuristic

	On next flip?	
	Heads	Tails
Usual response	✓	
Actual probability	50%	50%

_____
_____
_____
_____
_____
_____

Biases and heuristics are often used to make quick decisions; however, they may make humans prone to errors.

Confirmation bias: _____
_____
_____
_____

Confirmation bias in everyday life:
_____
_____
_____
_____
_____
_____

Heuristics: _____
_____
_____
_____

Availability heuristic: _____
_____
_____
_____
_____
_____
_____
_____

Representativeness heuristic: _____
_____
_____
_____
_____
_____

Representativeness heuristic in everyday life:
_____
_____
_____
_____
_____
_____
_____
_____
_____
_____

Module 5: Problem Solving & Intelligence – Courseware Exercise

Now that you've learned about problem solving, you can do a couple of actual tasks to demonstrate insight learning. Try your best at the following problems (and try not to look at the answers on the next page).

Wine problem

You have a glass of red wine and a glass of white wine, both with equal volume. You take exactly one teaspoon of red wine, put it into the white wine glass, and stir. Then, you take exactly one teaspoon of this new mixture and put it back into the red wine glass. In the end, both glasses have the same volume.

Question: Which glass is more pure (contains more of its original liquid)?

This is a tough problem and it may help if you give yourself simple values and to map it out on paper. If you manage to solve it in this way, you can see that giving yourself tangible values or a diagram really help in problem solving.

Now, try your hand at a second problem.

Tumour Problem

You are a doctor with a patient that has a malignant tumour in his stomach. To operate on the patient is impossible, but unless the tumour is destroyed, the patient will die. A kind of ray, at a sufficiently high intensity, can destroy the tumour. Unfortunately, at this intensity the healthy tissue on the way to the tumour will also be destroyed. At lower intensities, the rays are harmless to healthy tissue, but will not affect the tumour.

Question: Can the rays be used to destroy the tumour without injuring the healthy tissue?

This question may be a bit difficult and have you thinking that the solution lies in some obscure medical textbook. However, if we reframe the question with something practical, it becomes easier.

Tumour Problem Rephrased

You are raiding a castle and need 100 men attacking at once to win. Your men know how to construct a bridge that carries 10 men.

Question: How can you storm this castle?

The interesting thing is that if we present this question using a practical example like a castle with bridges, most people get it right. Problem solving is best with tangible things!

Solution to the Wine Problem

They actually are equally pure. To try to simplify it: suppose you have 100ml in each class, and you take 10ml out of the red and put it in the white. The white now was 100ml white, and 10ml red. If you take 10ml out again, 100/110 of that teaspoon will be White, and 10/110 will be Red, those are just the proportions in the glass. Basically, you take lots of red out of the red and put it in the white, then you take lots of white out of the white and put it in the red, putting some red back with it. In the end, both glasses will have lost the same amount; the white lost it all at once, but the red lost more, then got a bit back. Tricky, tricky.

Solution to the Tumour Problem

When you rephrase the problem in terms of a castle under attack, the answer is obviously to build 10 bridges! The easiest solution here is to use many of the low power rays from many directions such that they all meet and sum at the point of the tumour.

After solving these problems, do you think that problem solving ability is related to intelligence? How so?

_____

_____

_____

Obviously, these types of problems would be incredibly difficult to administer to animals. Is problem solving a uniquely human trait?

_____

_____

_____

How is problem solving different in animals? Can you think of a way to administer an insight-like problem to an animal?

_____

_____

_____

Module 5: Problem Solving & Intelligence – Review Questions

1) Describe the difference between deductive and inductive reasoning.
   a. How are these types of reasoning relevant for the scientific method?

2) What is the confirmation bias? Provide an example.

3) What is the availability heuristic? Provide an example.

4) What is the representativeness heuristic? Provide an example.

5) James' friends never open textbooks but consistently get A+ grades. James thinks this must be the best way to study and gets an F on his first biology midterm. Which cognitive short-cut did James use to make this decision? Explain.

6) What is the difference between reliability and validity?

7) Which of the following statement is the primary explanation of the Flynn Effect?
   a. Humans are becoming increasingly more educated, thus getting smarter.
   b. Due to the increased supply of nutritious foods, children today enjoy more rapid growth in their intellectual abilities.
   c. Humans are provided with more opportunities to be intellectually simulated, thus becoming more intelligent.
   d. None of the above.

8) Is IQ genetic, environmental, or both? Explain.

9) Suppose you ask Sam (6 years old) and Ella (9 years old) to tell you whether the following statement is correct. "If all squares are rectangles, and all rectangles are red, are all squares red?" Which of your children might Piaget expect to answer this question correctly?
   a. Sam
   b. Ella
   c. Neither children
   d. Both children

Module 5: Problem Solving & Intelligence – Quiz Question

Using the Wechsler scale of intelligence tests, which of the following is correct?

   a. Identical twins Sam and Simon showed very different IQ scores, suggesting genetic factors play an insignificant role on intelligence. (9%)
   b. Marge scores an IQ of 113, while her mother Sandra scored the same IQ when she was Marge's age. Marge thinks this supports her claim that she is more intelligent than her mother. (26%)
   c. The consistent mean IQ of 100 reflects how the overall raw scores from intelligence tests of the population have remained stable. (27%)
   d. Research on fraternal twins suggests a smaller role of the environment in determining one's intelligence. (38%)

This is a question from the 2011 Problem Solving and Intelligence Avenue Quiz that the majority of students struggled with. Together, we are going to identify the correct answer option and the reasons why each distractor is incorrect. In this example, only 26% of students selected the correct answer option, B. Let's go look at the options:

   a. This option provides us with a hypothetical situation in an attempt to prove a point, that genetic factors are insignificant when it comes to their influence on intelligence. If we think back to the web module for a second, a significant portion of time was devoted to explaining that both environment and genetics play important roles in one's intelligence, so this option must be incorrect. In fact, most students recognized this as an incorrect option.
   b. Once again, this option presents us with a hypothetical situation in which an IQ score is being compared to an IQ score received many years prior. It is always important to pull out the crucial information from a hypothetical answer option or question stem as this makes it much easier to identify whether the information contained within is correct or not. In this case, we know that for some strange reason (Flynn Effect) **raw** scores on the IQ test have been on a steady rise since the 1950's. Remember, an IQ score of 115 is standardized, which means the original raw score is compared to the raw scores of everyone else at the time of examination and then given a standardized score with a mean of 100. It is possible, therefore, that a standardized score of 115 today corresponds to a higher raw score than one associated with a 115 many years ago. Let's hold onto this option for now. (**Correct!**)
   c. This is a tricky option. Many times, question developers will use factual information in an inappropriate manner to prove a point. In this case, that a consistent, **standardized** mean IQ of 100 is evidence that the overall **raw** scores of the population have remained stable. It is true that the standardized mean of 100 always remains the same but is this really evidence to suggest that the raw scores also remain the same? For starters, we know that in order to receive a standardized score, a raw score must first be compared to everyone else taking the test at the same time. This fact alone suggests to us that it is possible that mean raw scores can change while the standardized mean score of 100 remains the same. On top of that, we know that there is an observation called the Flynn Effect in which raw scores have been on the rise since the early 1950s while the mean standardized score always remains 100. After considering all of this, this option does not seem correct.
   d. Let us go back to the question once again. It reads as follows: "**Using the Wechsler scale of intelligence tests**, which of the following is correct?". Any answer option we consider must have to do with the Wechsler scales of intelligence. When first examining this option, there is no direct

mention of an IQ test, only intelligence. This is our first red flag that this option may be incorrect, so we should continue cautiously. This option is trying to convince us that there is evidence to suggest a smaller role of the environment in determining intelligence. Smaller than what? This is another red flag. Let's assume the question writer is implying genetics. Was there ever information in the web module definitively stating that genes or the environment were more important in determining intelligence? No. There was evidence to suggest that each was important, but in no way was there evidence put forth to suggest that one was more important than the other. It is for this reason that we can eliminate this option.

This leaves us with **option B**, which is the correct response.

Key Terms

Arch of Knowledge	Functional Fixedness	Representativeness Heuristic
Availability Heuristic	Inductive Reasoning	Reversible Relationships
Concrete Operational Stage	Intelligence	Sensorimotor Stage
Confirmation Bias	Multiple Intelligences	Seriation
Conservation	Object Permanence	Stanford-Binet Intelligence Test
Deductive Reasoning	Piaget's Stages of Development	Validity
Flynn Effect	Preoperational Stage	Weschler Scales
Formal Operational Stage	Reliability	

Module 5: Problem Solving & Intelligence – Bottleneck Concepts

Piaget's Stages of Development
IQ Scores vs. the Flynn Effect
Inductive vs. Deductive Reasoning
Confirmation Bias

Inductive vs. Deductive Reasoning

Inductive reasoning takes place when you know a specific fact and create a generalization from that fact. For example, concluding that all dogs are vicious when you get bit by one involves inductive reasoning. This type of thinking is what often leads to stereotypes. Deductive reasoning, on the other hand, occurs when you know a general idea and you use it to predict specific facts that must be true if the general idea is true. For example, if you believe that studying diligently results in good marks, you may predict that your classmate, Fred, must have exceptional grades since you always see him studying at the library. These two concepts are often confused. The easiest way to differentiate them is to first identify the information you have (is it a general idea or a specific fact?) and then identify what you can hypothesize based on the information (is it a concrete conclusion or a general idea?). Let's try an example:

Marcus knows that when people don't get enough sleep, their mind does not function optimally. He notices one day that his friend Kathy did uncharacteristically bad on a midterm and concludes it must be because she didn't get enough sleep.

In this case, Mark starts off with the assumption that when people don't get enough sleep, their brains do not function as well. This is just a general idea about how the world works. He applies this fact to Kathy's situation. She didn't do as well as she normally does on a midterm and he concludes that it must be because she didn't get enough sleep and so her brain did not function as well during the midterm (a specific case). Since Mark goes from a general idea to a prediction, this is an example of deductive reasoning.

Test your Understanding

1) In the following scenarios, identify if inductive or deductive reasoning is being used:
a) Karen was surprised to find out that her friends Brian and Patricia started a relationship with each other. They were not alike whatsoever! This leads her to guess that maybe, in romantic relationships, opposites attract.

Answer: Karen is using inductive reasoning. She starts from a specific case/observation that her two polar-opposite friends, Brian and Patricia, began dating and concludes with a general idea that opposites attract in romantic relationships.

b) One day, Matthew notices that Wendy has come to class with a completely new outfit. He knows that Wendy only buys clothes when they are on sale. He concludes that there must be a big sale happening at the mall.

Answer: Matthew is using deductive reasoning. He starts with the general idea that Wendy only buys clothes when they go on sale, and concludes that there must be a sale at the mall (a specific fact) because she has a new outfit.

2) Come up with your own examples of inductive or deductive reasoning. Ask your TA or post it on the discussion boards to receive feedback on whether or not your example is correct.

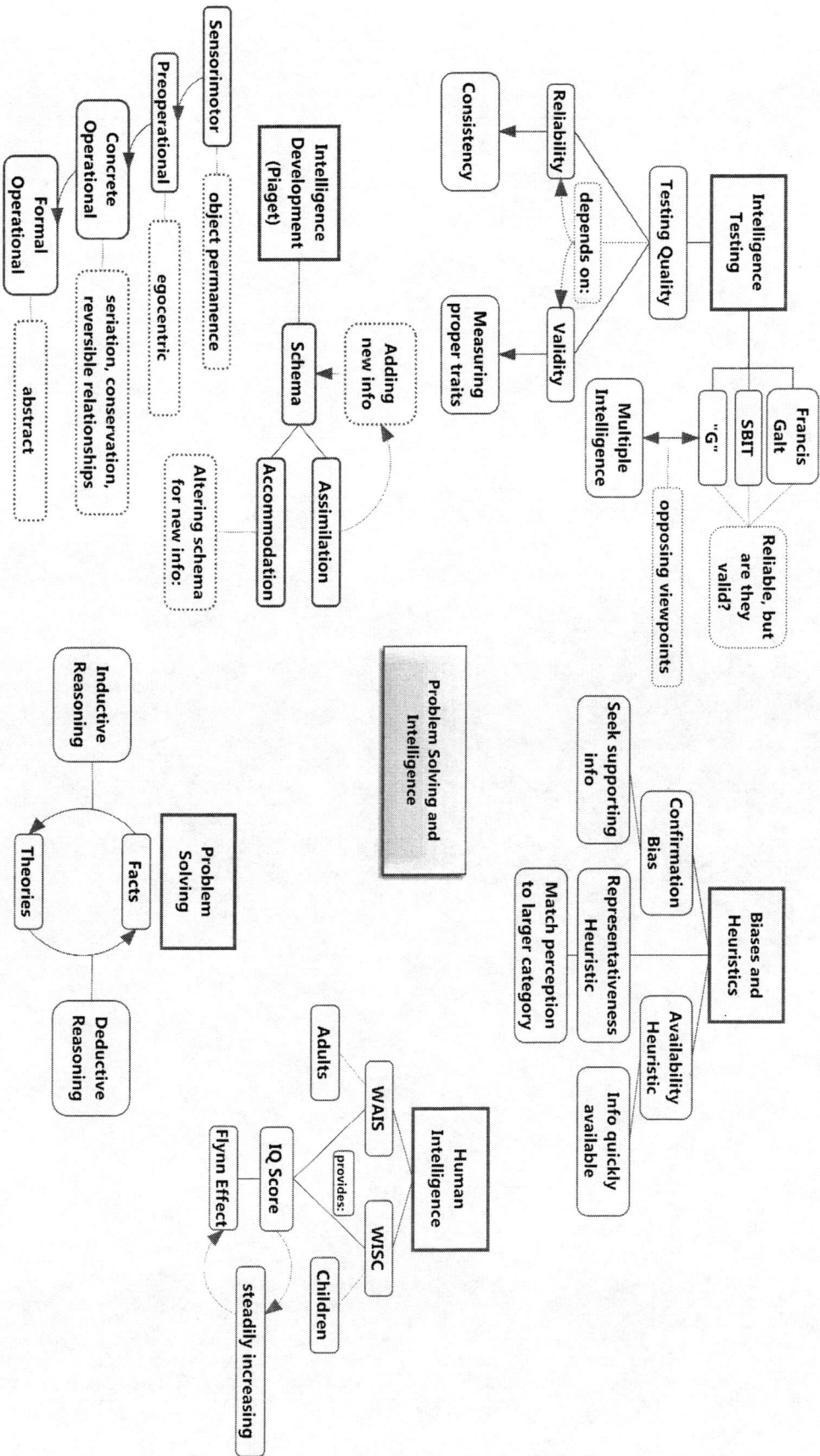

Intelligence Testing

- Testing Quality
  - depends on:
    - Reliability → Consistency
    - Validity → Measuring proper traits
  - Multiple Intelligence ←→ "G"
    - Francis Galt
    - SBIT
    - Reliable, but are they valid?
    - opposing viewpoints

Intelligence Development (Piaget)

- Sensorimotor → object permanence
- Preoperational → egocentric
- Concrete Operational → seriation, conservation, reversible relationships
- Formal Operational → abstract
- Schema
  - Adding new info
    - Assimilation
  - Altering schema for new info:
    - Accommodation

Problem Solving and Intelligence

Problem Solving

- Inductive Reasoning → Theories
- Facts
- Deductive Reasoning

Biases and Heuristics

- Confirmation Bias → Seek supporting info
- Representativeness Heuristic → Match perception to larger category
- Availability Heuristic → Info quickly available

Human Intelligence

- WAIS → Adults
- WISC → Children
- IQ Score (provides:)
  - Flynn Effect → steadily increasing

WEEK of Oct 13: LANGUAGE

"It depends on what the meaning of the word 'is' is."
- Bill Clinton, during his 1998 grand jury testimony on the Monica Lewinsky affair

Ellie and Janet are sitting in an art class working on their self-portrait painting project. Ellie, who is just beginning to explore her artistic side, turns to Janet looking for approval. Janet takes a look at Ellie's painting and casually says, "Actually, that's pretty." Ellie goes back to her work unsure what to make of the comment. Was this a compliment or an underhanded dig? The problem is the word "actually". Remove it from the sentence and it is clearly a compliment and social norms are upheld. But add that single word and an innocent comment can take on a negative connotation. The word "actually" implies a sense of surprise that what follows is not expected. Ellie can translate the comment to mean, "wow, usually what you do is not good, but this time, it's good." Although we are often casual in the language we use, you can probably think about a misinterpreted conversation or email that created a problem. Whether you are applying for a grant, filling out an application, or meeting with a potential client, language does matter. It's something that speechwriters, editors and advertisers think about very carefully. A single word or turn of a phrase can profoundly alter the meaning, impact, and even the legality of a sentence. Consider the sensational case of the Clinton-Lewinsky affair in which the question of whether Bill Clinton perjured himself before the grand jury depended on the definition of the word "is". Clinton rationalized that he wasn't lying when he said to his top aides that with respect to Monica Lewinsky, "there is nothing going on between us." According to Clinton: "It depends on what the meaning of the word 'is' is. If 'is' means is and never has been, that is not—that is one thing. If it means there is none, that was a completely true statement...Now, if someone had asked me on that day, are you having any kind of sexual relations with Ms. Lewinsky, that is, asked me a question in the present tense, I would have said no. And it would have been completely true." In returning to our example on the impact of the word "actually," add or delete it from a few other sentences and see how you feel about receiving the following comments:

- Those jeans actually look good on you.
- You actually don't look fat.
- That dinner was actually really good.

It's not just being nit-picky. Word choice actually is a matter of importance. Think about this the next time you are writing an important email.

Weekly Checklist:
- ☐ **Web Modules to watch: Language, Library Research**
- ☐ **Readings: Chapter 4**
- ☐ **AVE Quiz 6**

Upper Year Courses:
If you enjoyed the content in this week's module, consider taking the following upper year courses:
- PSYCH 3C03 Child Language Acquisition
- PSYCH 3PS3 Psycholinguistics Lab
- PSYCH 3UU3 Psychology of Language
- PSYCH 3YY3 Evolution of Communication
- PSYCH 4L03 Cognitive Neuroscience of Language

Module 6: Language – Outline

Unit 1: Introduction to Language

Natural Language: Regular

The quick brown fox jumps over the lazy dog

The lazy dog was jumped over by the quick brown fox

Natural Language: Arbitrary

- Lack of resemblance between words and their meaning

Cat
Neko
Gato
Chat
Katt
Felino
Katze

Natural Language: Productive

To classify a form of communication as language, several important criteria must be met.

Regular: _____

_____

_____

_____

_____

_____

_____

_____

_____

Arbitrary: _____

_____

_____

_____

_____

_____

_____

_____

_____

_____

_____

_____

_____

Productive: _____

_____

_____

_____

_____

_____

_____

_____

_____

_____

_____

_____

_____

The Whorf-Sapir Hypothesis

The Whorf-Sapir Hypothesis suggests that language has the power to influence our thoughts and perceptions.

<u>Evidence supporting the Whorf-Sapir Hypothesis:</u>

_____
_____
_____
_____
_____
_____
_____
_____

<u>Evidence against the Whorf-Sapir Hypothesis:</u>

_____
_____
_____
_____
_____
_____
_____
_____
_____
_____
_____
_____
_____
_____

Unit 2: The Structure of Language

Morphemes

Definition

Morpheme
- The smallest unit of sound the contains information
- Often a word, but some words contain multiple morphemes

_____
_____
_____
_____
_____
_____

Languages often contain multiple layers of structure to allow for effective communication.

<u>Morphemes:</u> _____

_____
_____
_____
_____
_____
_____
_____
_____
_____
_____
_____
_____
_____
_____
_____

Phonemes

Phonemes: _____
_____
_____
_____
_____

Variation between languages: _____
_____
_____
_____
_____
_____

Syntax and Semantics

Definition

Syntax
- The rules that govern how sentences are put together
- Also known as grammar

Syntax and semantics: _____
_____
_____
_____
_____
_____

Variation between languages: _____
_____
_____
_____
_____
_____

Unit 3: Development and the Segmentation Problem

Language Development

Babbling
- Characterized by drawn-out intense use of a variety of combinations of vowels and consonants
- May sound like a real sentence or question due to the use of inflection and rhythm in the production of the babble
- Combinations progress to become real words

Table 1: Milestones

12 wks	Makes cooing sounds
16 wks	Turns head towards voices
6 months	Imitates sounds
1 yr	Babbles
2 yrs	Uses 50-250 words; uses 2 word phrases
2.5 yrs	Vocabulary > 850 words

_____
_____
_____
_____
_____

The acquisition of language is a challenging process that occurs at different rates over many years.

The general course of language development: _____
_____
_____
_____
_____
_____

Language Explosion: _____
_____
_____
_____
_____
_____
_____
_____

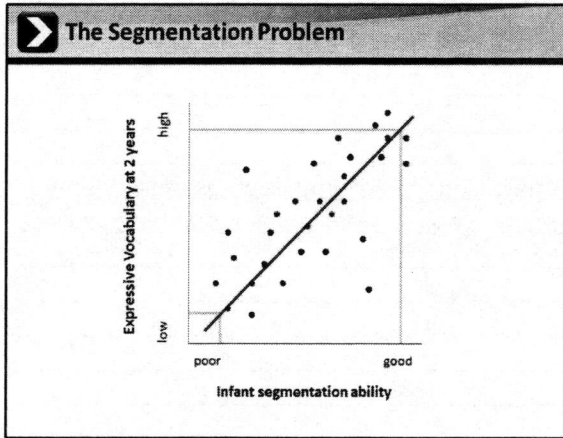

The Segmentation Problem

The segmentation problem: _____
_____
_____
_____

The relation of infant's segmentation ability to their later vocabulary: _____
_____
_____
_____
_____
_____
_____

Unit 4: Universal Phoneme Sensitivity

Universal Phoneme Sensitivity

Definition

Universal Phoneme Sensitivity
- The ability of infants to discriminate between any sounds they're tested on
- Includes sounds from non-native languages

Universal phonemic sensitivity: _____
_____
_____
_____
_____
_____
_____
_____
_____
_____
_____
_____
_____
_____
_____

Universal Phoneme Sensitivity

/ba/   /ɖa/

Methods of testing phoneme sensitivity in infants: _____
_____
_____
_____

Compare infants' phonemic sensitivity with that of adults: _____
_____
_____
_____
_____
_____

_____
_____
_____
_____
_____

_____
_____
_____
_____
_____
_____
_____

Universal Phoneme Sensitivity

The disappearance of universal phoneme sensitivity in infants: _____
_____
_____

Implications of varying degrees of phonemic sensitivity when learning a novel language: _____
_____
_____
_____
_____

Unit 5: Theories of Langage Development

Social Learning Theory

Two major theories of language development guide how researchers understand the acquisition of language.

Social Learning Theory: _____
_____
_____
_____
_____

Evidence supporting Social Learning Theory: _____
_____
_____
_____
_____

Overextensions and Underextensions

Evidence against Social Learning Theory: _____
_____

1) Overextensions: _____
_____
_____

2) Underextensions: _____
_____
_____
_____

_____
_____
_____
_____
_____
_____

_____
_____
_____
_____
_____

Innate Mechanism Theory

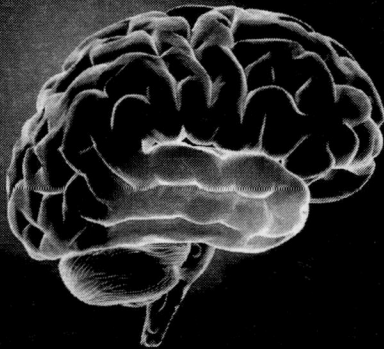

Language Acquisition Device
- An innate mechanism, present only in humans, that helps language develop rapidly according to universal rules

Innate Mechanism Theory

Unit 6: Animal Communication

The Waggle Dance

Waggle Phase
- Distance of waggle → Distance of food
- Angle of waggle → Direction of food

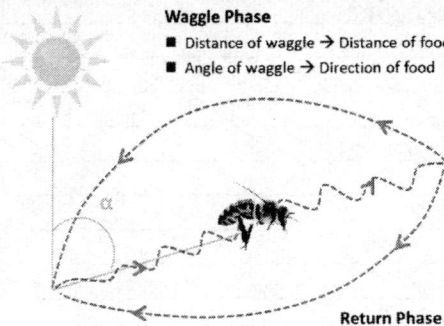

Return Phase

_____
_____
_____
_____
_____
_____

Innate Mechanism Theory: _____
_____
_____
_____
_____
_____

Noam Chomsky: _____
_____
_____
_____
_____
_____
_____

Evidence for the Innate Mechanism Theory: ___
1) Sign language: _____
_____
_____
_____
_____

2) Neural activation: _____
_____
_____
_____
_____
_____
_____

Human language can be compared to various methods of communication used by non-human animals.

Waggle dance: _____
_____
_____

Two phases: _____
_____
_____
_____

Duration: _____
_____
_____
_____
_____
_____
_____
_____
_____

Washoe

Washoe
- Could use signs to communicate requests
- Did not seem to use systematic grammar

Sarah

Sarah
- Taught to use symbols to communicate
- Used a large vocabulary, was able to answer questions
- Could not generate new sentences

Kanzi

Kanzi
- Taught to use lexigrams to communicate
- Utilized full immersion rather than classical conditioning

_____
_____
_____
_____
_____
_____
_____

Washoe: _____
_____
_____

Washoe's strengths: _____
_____
_____

Washoe's limitations: _____
_____
_____
_____

Sarah: _____
_____
_____

Sarah's strengths: _____
_____

Sarah's limitations: _____
_____
_____
_____

Kanzi: _____
_____
_____

Kanzi's strengths: _____
_____

Kanzi's limitations: _____
_____
_____
_____
_____
_____
_____
_____
_____
_____
_____

Module 6: Language – Courseware Exercise

Now that you have a rough idea of how cognitive researchers define language, it's important to apply these criteria to the animal communication examples that were covered in lecture and determine whether they exemplify true language.

Remember that language is *regular* in that it is governed by rules and grammar. Language is *arbitrary* because there is a lack of resemblance between a word and its meaning. Finally, language is *productive* because it can be recombined infinitely to describe countless new situations or ideas.

Use these criteria to evaluate each of the methods of communication in the following table. You may need to do a bit of online research to evaluate all three criteria for each example. If an example fulfills all three criteria, it can be considered language!

Animal example	Regular?	Arbitrary?	Productive?	Is it language?
Waggle Dance				
Bird Song				
Washoe				
Sarah				
Kanzi				
Human Speech				

Module 6: Language – Review Questions

1) Which of the following exemplifies the smallest unit of sound in the English language that contains information?
   a) The word "drivers"
   b) The letter "s" in the word "sock"
   c) The word "watch"
   d) The letter "y" in the word "party"

2) Which of the following situations best demonstrates the Whorf-Sapir Hypothesis?
   a) Infants born in an English-speaking family can discriminate phonemes presented in Vietnamese
   b) Adults who fluently speak the Inuit language have many words for "snow"
   c) Children who are not exposed to language from an early age still have the ability to spontaneously develop grammatical rules
   d) Animal communications are limited in their motor abilities, thus they can only convey messages relevant to their natural lifestyle

3) What are the milestones of infant language development?
   a. What is the language explosion?

4) What is the segmentation problem?
   a. How is an infant's ability to solve the segmentation problem related to their later language abilities?

5) What is universal phoneme sensitivity?
   a. Provide some evidence for the existence of this phenomenon.

6) Compare and contrast the two theories of language development

	Social Learning Theory	Innate Mechanism Theory
Supporting Evidence		
Opposing Evidence		

Module 6: Language – Quiz Question

What does the **language explosion** refer to?

   a. The peak of language development. (53%)
   b. Mastery of most linguistic components. (22%)
   c. Receptive vocabulary catches up with expressive vocabulary. (18%)
   d. Syntax is explicitly understood and used appropriately. (7%)

This is an actual question from the 2011 Language Avenue Quiz that many students struggled with. We are going to go through this question together to identify what sort of errors students were making and how to avoid them in the future. To do so, we will look at each answer option individually and identify why it is correct or incorrect.

To answer this question correctly, you must be reading each answer option carefully when comparing it to information from the web module. This question utilizes a number of answer options that, at first glance, seem correct but are, in fact, not. Let's have a look at the options.

   a. This is a very attractive option but, upon careful examination, is actually incorrect. Is the language explosion the peak of language development? The peak of language development suggests that language will develop no more as your language skills have been maxed out. When you reach the peak of a mountain there is no place to go but down. Does language development stop after the language explosion? No. In fact, mastery of syntax occurs long after the language explosion. Thus this option is incorrect.
   b. This seems like something we have seen in the web module and textbook verbatim. Let's keep this option. (**Correct!**)
   c. This seems like familiar textbook material. But we know that language explosion refers to expressive vocabulary. Considering this fact alone, this answer option appears incorrect.
   d. Is syntax explicitly understood? Do we cease to develop our understanding of syntax past the age of 6? No. We continue to master the syntax of language past the language explosion. Therefore, this answer option is also incorrect.

We are left with **option B**. It is often helpful to look for familiar wording. If a concept is worded in a bizarre way, this is often a clue that it is a nonsense answer option and should be eliminated.

Accent	Language Explosion	Social Learning Theory
Bird Song	Morpheme	Syntax
Foreign Accent Syndrome	Overextensions	Underextensions
Infant-Directed Speech	Phoneme	Universal Phoneme Sensitivity
Innate Mechanism Theory	Segmentation	Waggle Dance
Language	Segmentation Problem	Whorf-Sapir Hypothesis
Language Acquisition Device (LAD)	Semantics	

Module 6: Language – Bottleneck Concepts

Whorf-Sapir Hypothesis
Innate Mechanism Theory vs. Social Learning Theory
Universal Phoneme Sensitivity
Segmentation Problem

Whorf-Sapir Hypothesis
The Whorf-Sapir Hypothesis states that language influences our thoughts and the way we perceive and experience the world. If some words aren't present in a particular language, it may affect how people perceive the world if that is the only language they know. For example, in lecture, we discussed the Piraha, a group of Brazilians, and their language.. They only have three counting words: one, two, and many. The Whorf-Sapir hypothesis would predict that these people should have difficulty counting objects that are three or more. Consistent with this hypothesis, the Piraha had great difficulty making groups of objects as the groups became greater or equal to three. However, this hypothesis does not gain consistent support. For example, in Chinese, a separate word is used to describe an aunt and an uncle from your mother's side of the family versus your father's side of the family. Even though these words are not in the English language, people who only know English are still able to distinguish which side of the family their aunts and uncles belong to. Questions surrounding the Whorf-Sapir hypothesis will usually involve whether or not statements or scenarios agree with it. Always ask yourself if people's understanding of a concept is limited because their vocabulary lacks a word or enhanced due to the presence of a word; if either scenario is true,   the Whorf-Sapir hypothesis is supported. Below are a few examples related to this concept. Determine which ones lend support to or go against the Whorf-Sapir hypothesis.

Testing your Understanding
1) Which of the following statements **do not** support the Whorf-Sapir hypothesis?

a) In English, there are separate words for the colours, green and blue. The Tarahumara people on the other hand, have a single word that describes green/blue. English speakers are better at distinguishing green and blue than the Tarahumara people.
b) The Inuit language has more words describing snow compared to English. People who only know English have a harder time distinguishing between different types of snow.
c) In Greek, there many words for love including: Agápe (unconditional love), Éros (romantic love), Philia (friendship love), and Storge (family love). Even though Fred only knows English, he has no trouble recognizing that when his best friend Kyle says, "I love you man," he means it as a friend, and not in a romantic way.
d) In spoken Chinese, there is no distinction between he, she, or it. When people learn English as a second language, they have a hard time differentiating deciding whether to use he, she, or it in sentences.

Answer: C. Even though the English language does not contain different words for each type of love, Fred is still able to identify which type of love his friend is referring to. This example does not support the Whorf-Sapir hypothesis which would argue that a lack of different words for love would have hindered Fred from differentiating between the various types of love. All other choices support the Whorf-Sapir hypothesis since the absence of a word in an individual's language altered the way in which they perceived the world.

2) Can you name other examples where words are present in one language but not another? In those cases, do you think having that extra word influences how people perceive or experience the world? Does this example support or go against the Whorf-Sapir Hypothesis?

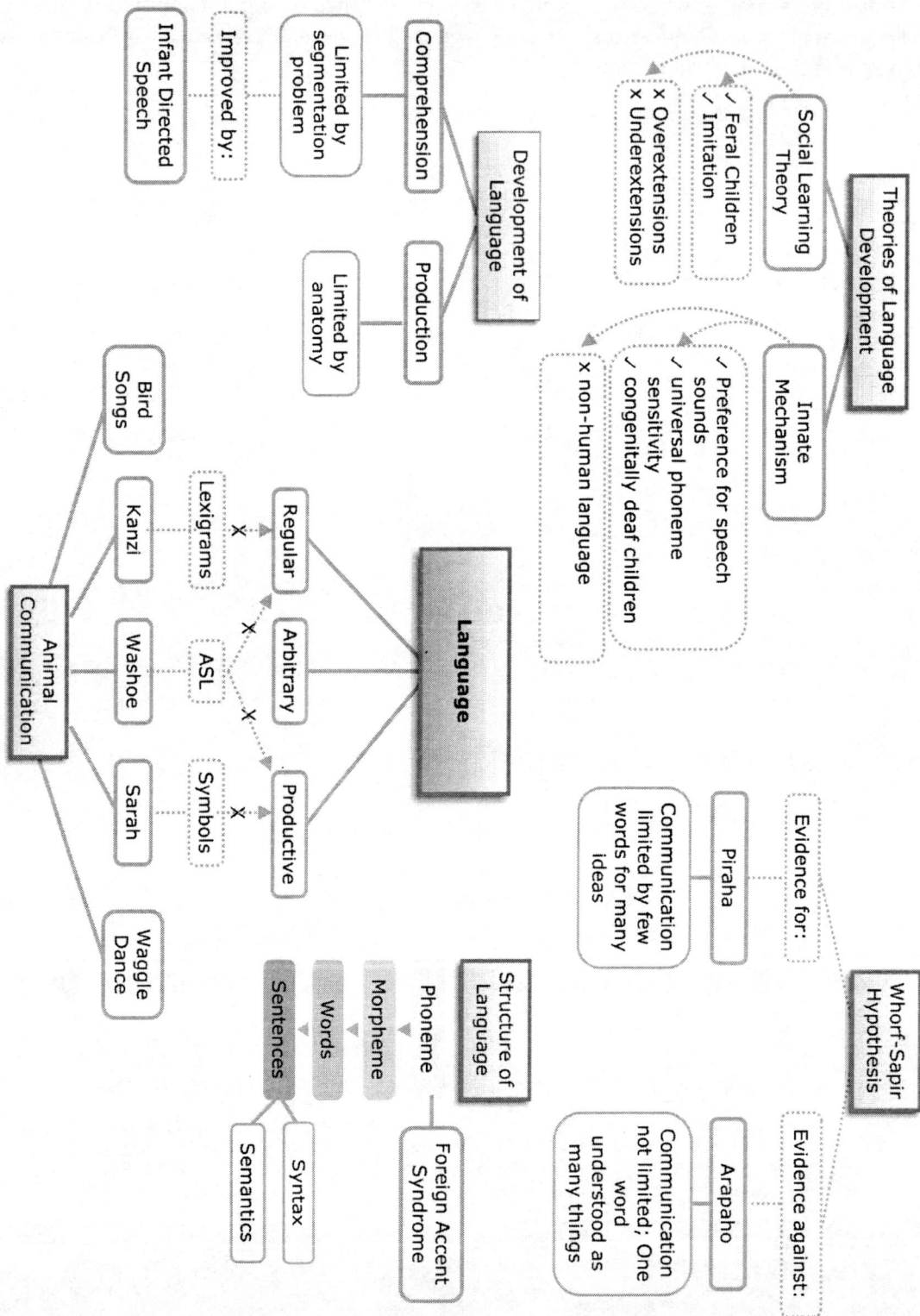

Development of Language

Comprehension
- Limited by segmentation problem
  - Improved by:
    - Infant Directed Speech

Production
- Limited by anatomy

Theories of Language Development

Social Learning Theory
- ✓ Feral Children
- ✓ Imitation
- ✗ Overextensions
- ✗ Underextensions

Innate Mechanism
- ✓ Preference for speech sounds
- ✓ universal phoneme sensitivity
- ✓ congenitally deaf children
- ✗ non-human language

Language

Animal Communication
- Bird Songs
- Kanzi — Lexigrams
- Washoe — ASL
- Sarah — Symbols
- Waggle Dance

Regular ✗
Arbitrary ✗
Productive ✗

Whorf-Sapir Hypothesis

Evidence for:
- Piraha — Communication limited by few words for many ideas

Evidence against:
- Arapaho — Communication not limited; One word understood as many things

Structure of Language
- Sentences
- Words
- Morpheme
- Phoneme
  - Foreign Accent Syndrome

Sentences:
- Semantics
- Syntax

WEEK of Oct 20: CATEGORIES & CONCEPTS

"What's in a name? That which we call a rose by any other name would smell as sweet."
- Shakespeare, Romeo and Juliet

The label used to identify a disease—whether it is common language or medical terminology— can influence how serious people think the condition is. For example, impotence is now widely known as erectile dysfunction; excessive sweatiness is also known as hyperhidrosis. When people are presented with the medicalized term for these recently medicalized conditions, they are perceived to be more severe, more likely to be a disease, and more likely to be rare compared to the same disorder presented with its synonymous lay label. Participants in a study were given a survey that included 16 disorders, eight of which were chosen due to the increased popular use of a medical label within the last 10 years (e.g. erectile dysfunction versus impotence). The remaining eight were established medical disorders with both lay and medical terminology in popular use for more than 10 years (e.g. hypertension versus high blood pressure). "A simple switch in terminology can result in a real bias in perception," says Meredith Young, one of the study's lead authors and a graduate student in the Department of Psychology, Neuroscience & Behaviour at McMaster University." These findings have implications for many areas, including medical communication with the public, corporate advertising, and public policy." The pattern of results also has direct implications for the patient. If a patient is informed that she has gastroesophageal reflux disease, for example, rather than chronic heartburn, she might think she is more ill. An important implication is that the patient's understanding of the condition heavily influences how she goes about taking care of her own health. Pharmaceutical companies may have an inherent interest in the medicalization of disease. The more serious a condition is perceived to be, the more convinced you may be that a drug treatment is appropriate. And so, while a red rose or rosacea may smell just as sweet, they may be perceived very differently.

Young, M. E., Norman, G. R., & Humphreys, K. R. (2008). Medicine in the popular press: the influence of the media on perceptions of disease. *PLoS One, 3*(10), e3552.

Weekly Checklist:
- ☐ **Web Module to watch: Categories & Concepts**
- ☐ **Readings: Journal Article**
- ☐ **AVE Quiz 7**

Upper Year Courses:
If you enjoyed the content in this week's module, consider taking the following upper year course:
- PSYCH 2E03 Sensory Processes

Module 7: Categories & Concepts – Outline

Unit 1: Categorization

Categorization allows us to deal with our surroundings quickly and effectively.

_____
_____
_____
_____
_____
_____
_____
_____
_____
_____

Unit 2: Functions of Categorization

Categorization serves four specific functions.

1) Treating non-identical objects in a similar manner: _____
_____
_____
_____
_____

2) Understanding our surroundings: ___
_____
_____
_____
_____
_____

3) Using current experience to predict future events: _____
_____
_____
_____

4) Communication: _____
_____
_____
_____
_____
_____

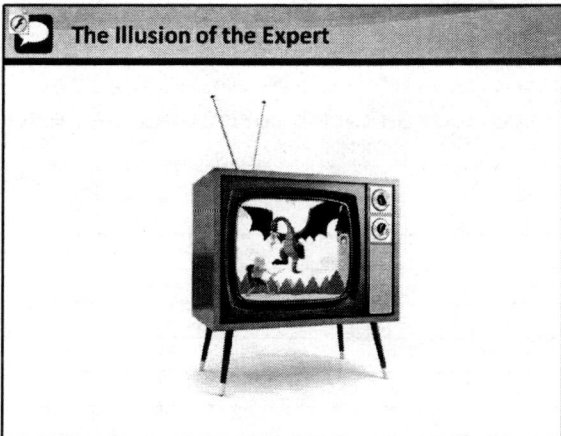

The Illusion of the Expert

Categorization seems to be a simple process because humans can perform it with great ease.

Illusion of the expert: _____
_____
_____
_____
_____
_____
_____
_____
_____

Unit 3: Rules

Dr. Lee Brooks

%yes	Category	Include...	But not...
84%	table	co...	...d, counter, iron board
88%	bottle	ba...	jar, glass, carton
64%		Chihuahua, greyhound	...oyote
84%		sapling, pine, palm, bonsai	bus... ...bamboo
40%	...uit	melon, coconut, grape	
56%	furniture	rug, chair, desk	table saw

...and those are the categorization rules for the perfect girl to date.

Rules are often insufficient to describe all members of a category.

Research by Dr. Brooks: _____
_____
_____
_____
_____
_____
_____
_____
_____
_____
_____

When Rules Aren't Enough

_____
_____
_____
_____
_____
_____

Implications of humans' inability to articulate rules for various categories: _____
_____
_____
_____
_____
_____
_____
_____
_____
_____
_____
_____
_____
_____
_____
_____
_____

Unit 4: Prototype Theory

Prototypes

fruit   bird

Prototype theory suggests that we categorize our current experience by comparing it to average representations of various categories.

_____
_____
_____
_____

The role of experience: _____
_____
_____
_____
_____

Categorization Using Prototypes

Evidence supporting the prototype theory: _____
_____
_____
_____
_____

Evidence against the prototype theory: _____
_____
_____
_____
_____
_____

Unit 5: Exemplar Theory

Exemplars

_____
_____
_____
_____
_____
_____

Exemplar theory suggests that our entire lifetime of experiences are stored and used to categorize our current experiences.

_____
_____
_____
_____
_____
_____
_____
_____
_____
_____
_____
_____
_____
_____
_____

Prototype vs Exemplar Theory

Use the exemplar theory to explain how robins can be classified as birds more quickly than penguins: _____
_____
_____
_____

The dermatologist experiment: _____
_____
_____
_____
_____
_____

Unit 6: The Development of Categorization

Children and Categories

Children begin developing their categorization skills during the first few years of life.
_____
_____
_____
_____

Children's innate understanding of categories: _____
_____
_____
_____
_____
_____
_____

Unit 7: Animal Categorization

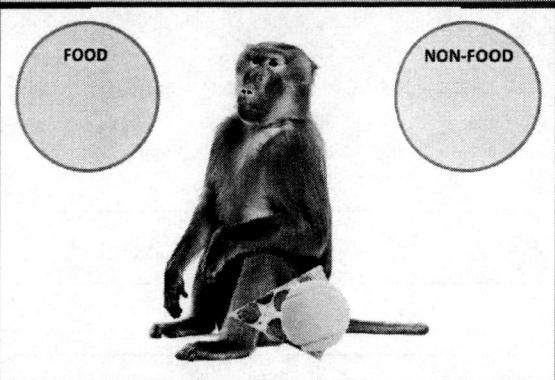

Baboon Categorization

FOOD   NON-FOOD

Some animal species can become highly skilled at categorization.
Food versus non-food: _____
_____
_____
_____

Same versus different: _____
_____
_____
_____
_____
_____
_____
_____
_____

_____
_____
_____
_____
_____

128

Module 7: Categories & Concepts – Courseware Exercise

A researcher is measuring typicality and group membership ratings for a variety of objects. To determine typicality ratings, she asks participants to rate how typical objects are to their greater category (i.e. how well does "cow" *represent* the category "barnyard animals"?). She then asks participants for the membership ratings of the same objects (i.e. how well does "cow" *belong to* the category of "barnyard animals"?).

She found the following results for the objects "maple" and "palm" for the category "trees": palm trees were found to be non-typical, but were generally thought to belong in the category "trees". Maple trees were rated high on both typicality and membership.

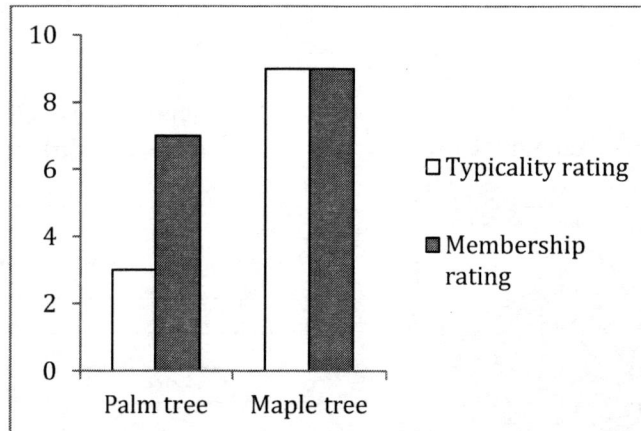

If you were told that the participants in this experiment had lived their entire lives in Canada, how would you explain the discrepancy between typicality and membership ratings for the object "palm tree"?

_____

_____

_____

Which of these two trees would be regarded as the ideal member of the category "trees"? What piece of data from the study supports your answer?

_____

_____

_____

Could the results of this study be explained using exemplar theory? Explain how.

_____

_____

Module 7: Categories & Concepts – Review Questions

1) Provide your own example to explain each of the four advantages of categorization.

2) Is it possible to describe every category with a set of rules?
   a. What do experiments on the use of rules tell us about categorization?

3) Compare and contrast the prototype theory with the exemplar theory.

	Prototype Theory	Exemplar Theory
Similarities		
Differences		
How does the theory explain the categorization of an "orange" into the category "fruit"?		

4) Describe the development of categories and concepts.
   a. List evidence that provides support for the idea that children can apply categories.

5) List evidence that provides support for the idea that animals can apply categories and concepts?

Module 7: Categories & Concepts – Quiz Question

The results of the study conducted by Rips (1973) on the reaction time needed to categorize a penguin and a robin as birds can be explained by both prototype and exemplar theories. Which of the following statements is accurate?

a. According to **prototype** theory, it should take longer to classify a robin as a bird because the average representation of the category *bird* is likely closer to a robin than a penguin. (4%)

b. According to **exemplar** theory, *robin* should be matched to existing experiences of *bird* more quickly than *penguin* because there are likely many more robin exemplars. (75%)

c. According to **prototype** theory, *robin* should be matched to existing experiences of *bird* more quickly than *penguin* because there are likely many more *robin* exemplars. (19%)

d. According to **exemplar** theory, it should take longer to classify a robin as a bird because the average representation of the category *bird* is likely closer to a robin than a penguin. (2%)

This is a question from the Categories & Concepts Avenue Quiz in 2011. This question asks students to apply their knowledge about exemplar and prototype theories to the Rips (1973) study. When approaching this question, it is be beneficial to consider what was presented in the web module about this study and about each theory of categorization. Let's look at our options:

a. This option uses a common question writing technique where a portion of the answer is correct but one aspect is incorrect. It is true that prototype theory considers the average representation of a bird and this average or best example is probably closer to a robin than a penguin for people in our society. However, because this is true, it should also take longer to classify a *penguin* as a bird, whereas this option says that it should take longer to classify a robin. Therefore, this option is incorrect.

b. This option correctly describes categorization by exemplar theory as comparing new experiences with a bird to existing experiences (exemplars) of birds. Because people in our society likely have many more experiences with robin-like birds, than with penguins, they should be faster to categorize a robin as a bird. This answer seems to be correct. (**Correct!**)

c. This option correctly states that robins are likely to be categorized faster than penguins; however, it describes prototype theory as using previous experiences or exemplars to categorize. Therefore, we know option C is incorrect.

d. There is no correct portion to this option. We know from the study discussed in the web module that robins are categorized as birds faster than penguins. We also know that prototype theory uses comparison to an average or best example of a category member, while exemplar theory uses comparisons to all previous experiences with category members to categorize. Therefore, we can rule out option D when selecting the best choice.

This leaves us with **option B** as the correct choice!

Module 7: Categories and Concepts – Bottleneck Concepts

Exemplar vs. Prototype theory
Rips' (1973) experiment
Brooks et al. (1991) experiment with skin experts
Animal Categorization

Brooks et al. (1991) experiment with skin experts

Understanding the results of experiments and how they support or contradict certain theories is an excellent way to test your understanding of course content. Examining the experiment performed by Brooks et al. in 1991 will help you better differentiate between the exemplar and prototype theories. In this experiment, experienced dermatologists were asked to diagnose various skin diseases after viewing their pictures on slides. Diagnosing a disease requires an individual to categorize the patient's symptoms as belonging to, and characteristic of, that disease. After 2 weeks, the same dermatologists returned to perform the task again. This time around, the slides were different but some of them depicted the same disorders presented during the previous session. The researchers wanted to determine if being exposure to certain skin disorders during the first week would help dermatologists better diagnose the same disorders two weeks later.

According to the prototype theory, examples from the previous week should not affect dermatologists' accuracy because recent example would simply be averaged with the hundreds of previously seen cases of the disorders. Therefore, the first phase of the experiment would cause a minimal change in the doctors' prototype for certain disorders, resulting in virtually no improvement in their accuracy.

On the other hand, the exemplar theory argues that recent examples would enable dermatologists to be more accurate in their diagnosis. The exemplar theory can be thought of as a file cabinet where each example of a category is stored. Everything you try to categorize is compared to each example until a similar match is found. Recent examples will be compared first so any recent examples of a certain disorder would help dermatologists better and more quickly diagnose skin disorders. Adding support to the exemplar theory, researchers found that skin experts were 20% more accurate in diagnosing the disorders they had been tested on two weeks earlier. This experiment also highlights the process of scientific research. Investigators can use evidence from this study to argue that certain categorization tasks are performed in the manner outlined by the exemplar theory.

Test your Understanding

1) What would be the results of this experiment if they had supported the prototype theory instead of the exemplar theory?

Answer:
In prototype theory, all relevant examples of a category are averaged into a prototypical depiction which represents the group. All future items are categorized by comparing them to prototypes. Since hundreds of examples contribute to a category's prototype, a few recent examples are unlikely to change the prototype significantly. Therefore, an improvement in the dermatologists' ability to diagnose recently viewed skin diseases should not occur.

2) Although people differ slightly in how they say certain words or phrases, we are able to understand their

speech. Categorization may play a role in this phenomenon. This difference is more noticeable when visiting another country. When Frederick was 25, he travelled to Australia for 2 weeks. Their accent seemed strange to him at first, but he eventually became accustomed to it. When he returned home from his vacation, Frederick thought that people in his hometown also had a strange accent that he hadn't noticed before. Does Fredrick's case support the exemplar or the prototype theory? Explain.

Answer:

This phenomenon supports the exemplar theory. Upon his return home, Frederick seemed to think that people in his hometown had a funny accent. Exemplar theory argues that all examples of a category are stored. Therefore, after hearing so many recent examples of the Australian accent, Frederick would first compare what he heard to the Australian accent when trying to categorize what his townspeople were saying. This would make it harder for Frederick to categorize the word, and thus make it seem like the townspeople had a funny accent. On the other hand, in the prototype theory, the recent examples of words said in an Australian accent should be averaged out with a lifetime of examples without that accent. Thus, this new experience would have been unlikely to change the prototype and affect how Frederick categorizes words heard back home if the prototype theory applied.

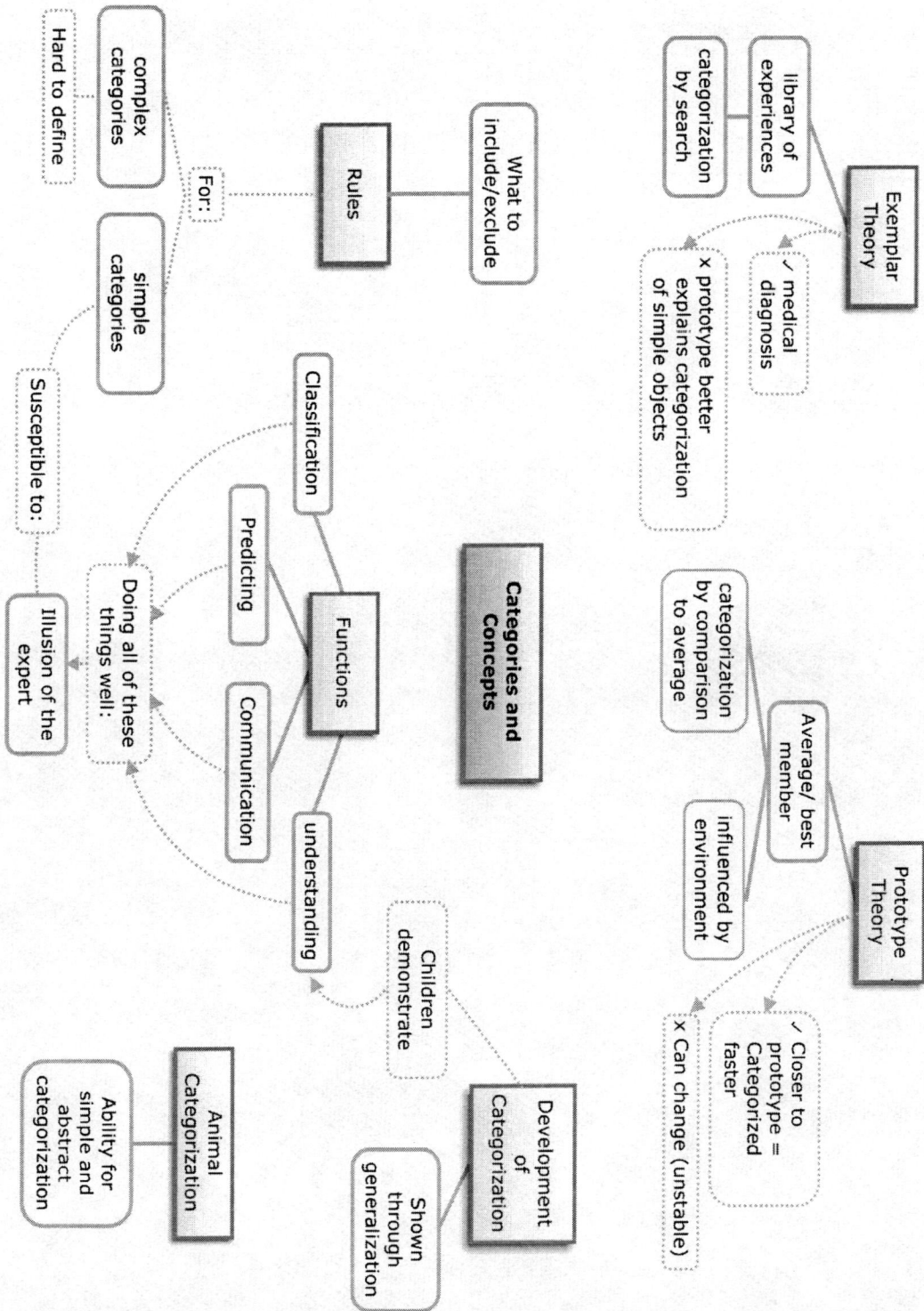

WEEK of Nov 3: ATTENTION

"Productivity in 11 words — one thing at a time. Most important thing first. Start now."
- Anonymous

Chances are that you are attending to many other things while reading this article. You may also be listening to music, surfing the web, or messaging friends via phone or email. Although attending to multiple tasks at the same time may make one feel both efficient and productive, there is much psychological research that indicates that our ability to engage in simultaneous tasks is, at best, limited and at worst, virtually impossible. The implications of these findings are significant, especially when we consider the use of technology in the classroom setting. With the ever increasing speed of information technology and the increasing use of laptops in the classroom, it is quite possible to be listening to one of the world's brightest minds, while also checking your email and reading the news at the same time. Given that the controlled processes involved in conscious attention are limited, are you really absorbing any information presented in a lecture while engaging in simultaneous tasks that require your attention? Hembrooke and Gay investigated the effects of multitasking in the classroom. In their experiment, two groups of students heard the exact same lecture and were tested immediately following it. One group of students was allowed to use their laptops to engage in browsing, search, and social computing behaviours during the lecture. In the other condition, students were asked to keep their laptops closed for the duration of the lecture. Results of the study demonstrated that students in the open laptop condition, who engaged in multiple tasks throughout the lecture, performed significantly worse on traditional measures of memory for lecture content. These results are hardly surprising, and by extension similar studies find that multitasking inhibits performance in a variety of occupational settings as well. The conclusion? Attention is the triumph over chaos.

Hembrooke, E., & Gay, G. (2003). The Laptop and the Lecture: The Effects of Multitasking in Learning Environments. *Journal of Computing in Higher Education, 15*(1), 46-64.

Weekly Checklist:
- ☐ **Web Module to watch: Attention**
- ☐ **Readings: Chapter 5 (1-4)**
- ☐ **AVE Quiz 8**

Upper Year Courses:
If you enjoyed the content in this week's module, consider taking the following upper year courses:
- PSYCH 2H03 Human Learning and Cognition
- PSYCH 3II3 Cognitive Development
- PSYCH 2E03 Sensory Processes

Module 8: Attention – Outline

Unit 1: Introduction to Attention

> William James

"Everyon[...] the taking pos[...] m, of one out [...] ously possible [...] mplies withdraw[...] o deal effectivel[...] n which has a [...] azed,

William James

> Selection

The concept of attention encompasses a broad array of topics that can be difficult to define.

William James on attention: _____
_____
_____
_____
_____
_____
_____
_____
_____
_____

Selecting information to attend to: _____
_____
_____
_____
_____
_____

The role of irrelevant stimuli: _____
_____
_____
_____
_____
_____

Unit 2: Automatic and Controlled Attention

AA Introduction

Definition

Automatic Processes
- Involuntary "capture"
- Fast, efficient

Controlled Processes
- Conscious attention
- Slow, effortful

Two essential processes determine where attention is directed.

Automatic processes: _____
_____
_____
_____
_____

Controlled processes: _____
_____
_____
_____
_____

Unit 3: The Spotlight Model

Selection and the Spotlight

Objects within the spotlight:
- Faster reaction time
- Higher accuracy

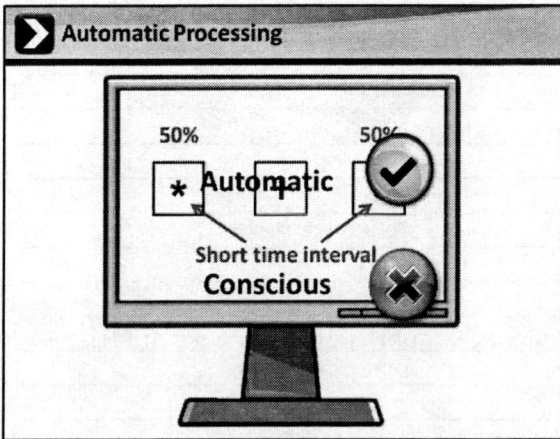

Automatic Processing

50%   50%

* Automatic

Short time interval

Conscious

Unit 4: Filter Models

Filter Models

_____
_____
_____
_____
_____

Psychologists use models to gain a well-rounded understanding of attention.

The Spotlight Model: _____
_____
_____
_____
_____
_____
_____
_____
_____
_____

Cueing paradigms can be used to measure changes in attention.

Target appears in flashed box: _____
_____
_____
_____
_____

Target appears outside of flashed box:
_____
_____
_____
_____

The Cocktail Party Effect: _____
_____
_____
_____
_____

Filter Model: _____
_____
_____
_____
_____
_____
_____
_____
_____
_____
_____

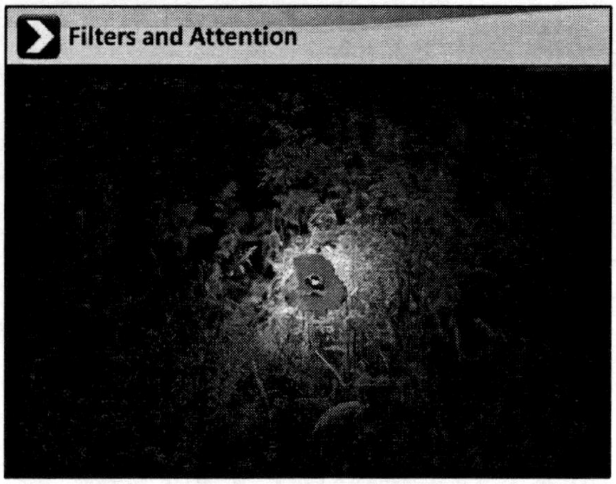

Filters and Attention

Broadbent's Single Filter Model

Attended Ear

Unattended Ear

Subjects remember nothing about unattended information

Broadbent's Single Filter Model

Target Ear

Distracter Ear

Von Wright et al.: Subjects show a response to unattended information

Difference between the spotlight and filter models:

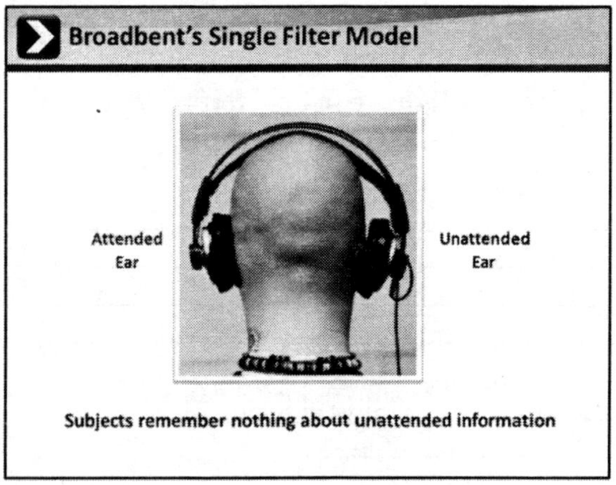

Broadbent's Single Filter Model:

Dichotic listening paradigm:

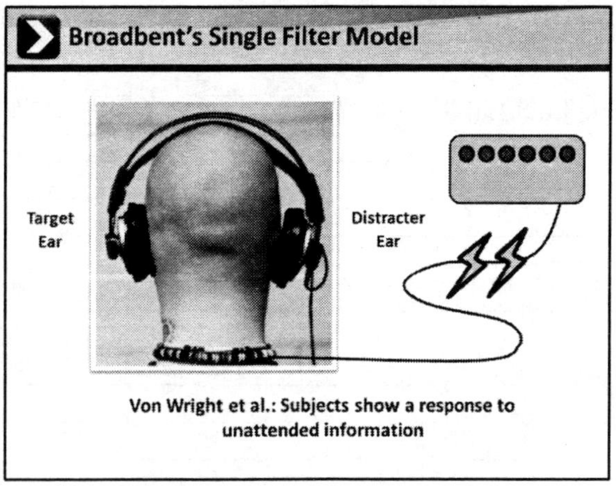

Evidence supporting Broadbent's model:

Limitations of Broadbent's model:

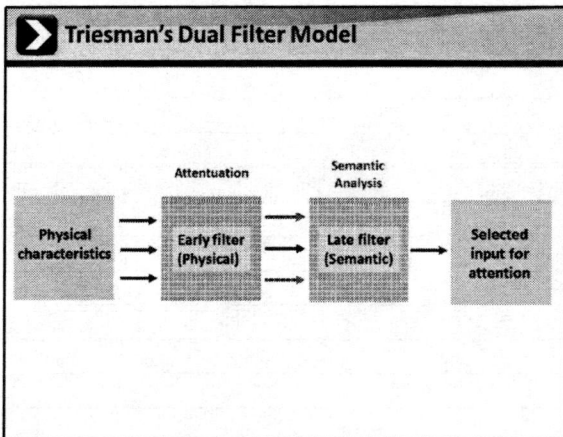

Triesman's Dual Filter Model

Triesman's Dual Filter Model:

_____
_____
_____
_____
_____
_____
_____
_____
_____
_____
_____
_____

Unit 5: The Stroop Task

Example Task

The Stroop Effect is a popular phenomenon that demonstrates the limitations of our attentional processes.

Congruent stimuli: _____
_____

Incongruent stimuli: _____
_____

Typical results: _____
_____
_____
_____

Controlling the Stroop Effect

Definition

Proportion Congruent Manipulation
- Change the ratio of congruent to incongruent trials

75% Congruent
25% Incongruent → Increased Stroop effect

25% Congruent
75% Incongruent → Decreased Stroop effect

Using the Stroop Effect to measure processes associated with attention:
_____
_____
_____

Results from various proportion congruent manipulations:
_____
_____
_____
_____
_____
_____
_____
_____
_____
_____

_____
_____
_____
_____
_____
_____

Controlling the Stroop Effect

(Congruent)

RED

> RED! Yes, I got it right!

(Incongruent)

BLUE

> Hm, it reads "blue"...

The Stroop Task: Automatic & Controlled

Automatic

Evidence: Word reading influences performance even when the word is to be ignored

Controlled

Evidence: People can adopt consciously controlled word reading strategies that modulate the Stroop effect

---

Unit 6: Visual Search

Visual Search Tasks

Definition

Set Size
- The number of items to search through

Set Size Effect
- Increase in difficulty as set size increases

_____
_____
_____
_____
_____
_____

Describe how consciously adopted strategies can influence the Stroop Effect:
_____
_____
_____
_____
_____
_____
_____
_____
_____

Describe how the Stroop Task can inform us about automatic and controlled processes:
_____
_____
_____
_____
_____
_____
_____
_____
_____
_____
_____

Visual search tasks mimic everyday tasks that require attention.

Set size:
_____
_____
_____
_____

Set size effect:
_____
_____
_____
_____
_____
_____
_____
_____

Feature and Conjunction Search

Definition

Pop-out Effect
- When the object of a visual search is easily found, regardless of set size.
- Easily induced by colour.

Contextual Cueing

Unit 7: Conclusion

The Importance of Attention

_____
_____
_____
_____
_____
_____

Single feature search: _____
_____
_____

Conjunction search: _____
_____
_____
_____
_____

Pop-out effect: _____
_____
_____

Describe the role of context in everyday search tasks: _____
_____
_____
_____
_____
_____
_____
_____
_____
_____
_____
_____

Conclusion
_____
_____
_____
_____
_____
_____
_____
_____
_____
_____
_____
_____
_____
_____
_____
_____
_____
_____

Module 8: Attention – Courseware Exercise

Debbie was driving down busy Main Street to work one morning. As she traveled, she tried to keep her attention focused on the road, but a red Thunderbird convertible, a few lanes over, suddenly caught her attention. She lost focus and began to veer into the next lane. Despite the fact that she had already passed several other red cars that morning, this one caught her eye since it was the same car her father drove.

Can this situation be explained using the spotlight model of attention? Explain why or why not.

_____

_____

_____

_____

Can this situation be explained using Broadbent's single filter model of attention? Explain why or why not.

_____

_____

_____

_____

Can this situation be explained using Triesman's dual filter model of attention? Explain why or why not.

_____

_____

_____

_____

Module 8: Attention – Review Questions

1) Describe the difference between automatic and controlled processes of attention.
   a. What is the advantage of using an automatic process?
   b. What is the advantage of using a controlled process?

2) Describe the spotlight model of attention.

3) Describe the spatial cueing paradigm.

   Answer the following questions regarding the spatial cueing paradigm:

   a. In Experiment #1, the target appears in the cued location for 26/30 trials. How will this affect response times? What type of attention does this demonstrate?

   b. In Experiment #2, the target appears opposite the cued location for 26/30 trials. How will this affect response times? What type of attention does this demonstrate?

   c. In Experiment #3, the target appears in the cued location for 16/30 of trials. How will this affect response times? What type of attention does this demonstrate?

4) Describe the difference between Broadbent's early filter theory and Triesman's early and late filter theory.
   a. What is breakthrough? Why was it important for the development of attention theory?

5) Describe the basic Stroop attention task.
   a. What does the Stroop task tell us about attention?

6) Describe a basic visual search experiment.
   a. What is the difference between feature and conjunction search?
   b. How is the pop-out effect related to the set size effect?
   c. Describe the relevance of contextual cueing.

Module 8: Attention – Quiz Question

In which of the following would it take participants the longest to find the target?

a. If participants are required to find a red triangle among 150 red squares. (33%)
b. If participants are required to find a red triangle among 70 purple and blue triangles. (3%)
c. If participants are required to find a blue square among 70 blue triangles and green squares. (57%)
d. If participants are required to find a green circle among 150 blue and yellow circles. (7%)

This is an actual question from the 2011 Memory & Attention Avenue Quiz that many students struggled with. Fortunately, we are going to go through this question together to identify what sort of errors students were making and how to avoid them in the future. To do so, we will look at each answer option individually and identify why it is correct or incorrect.

In this question, students were asked to apply the concepts they learned about visual search tasks to determine which scenario would be the most difficult search and, thus, take the longest to complete.

a. This option was a popular choice, as participants must search through 150 items to find the target, as opposed to only 70 in option C. However, this option describes a single feature search, where participants are only looking for a triangle in a field of squares. As was described in the web module, conjunction searches are more difficult than single feature searches and thus take longer to complete. Therefore, this option is incorrect.
b. It is easy to determine that this option is incorrect because of the popout effect. The target red triangle is the only red item in the field, which would make it easier to spot.
c. This option describes a conjunction search, where participants not only have to look for a blue target in a field of blue and green, but also a square in a field of squares and triangles. As was described in the web module, conjunction searches are more difficult than single feature searches, taking longer to complete. Thus, this is the option is the best choice. (**Correct!**)
d. It is easy to determine that this option is incorrect because of the popout effect. The target in this option is the only green item in the field, allowing it to "pop out" even in a field of 150 items.

This leaves us with **option C** as the correct answer.

Key Terms

Automatic Processes	Feature Search	Spotlight Model
Broadbent's Single Filter Model	Filter Model	Stroop Task
Conjunction Search	Pop-Out Effect	Triesman's Dual Filter Model
Contextual Cueing	Set Size	
Controlled Processes	Set Size Effect	

Module 8: Attention – Bottleneck Concepts

Spotlight vs. Filter Models
Single Filter Model vs. Dual Filter Model
Spatial Cueing Paradigm
The Stroop Effect

The Stroop Effect

The Stroop Effect was discovered in an experiment where participants were presented with a list of colour words written in different coloured inks. Their task consisted of naming the colour of the ink that the words were written in while ignoring the actual word that the letters spelt out. Two types of stimuli were presented throughout this experiment: (1) Congruent words, where the colour of the word matched the word itself, and (2) Incongruent words, where the colour of the word differed from the word itself. The Stroop Effect describes the observation that people take longer to name the colour of the ink in incongruent words than in congruent words. This is because in the case of incongruent words, the word itself intereferes with the colour-naming task while in congruent words the actual word facilitates the colour-naming task. Students often find it difficult to understand the effects of manipulating the proportion of congruent and incongruent words in a block of trials. When a high proportion of congruent words and a low proportion of incongruent words are presented, the Stroop Effect is increased — the participant requires more time than usual to name the ink color of incongruent words. This occurs because a high proportion of congruent words lead participants to read the word itself to complete the colour-naming task. However, since participants are consciously paying attention to the words, when an incongruent word appears, it especially hinders performance on naming the ink colour of the word.

A useful analogy to understand the Stroop Effect can be drawn from examining a driver's interaction with the GPS system in his car . If an individual always relies on the GPS to navigate to his final destination, he is more likely to get lost if the GPS fails. Similarly, if there are more congruent words, and the participant relies on the word itself to name the ink colour, then they are more likely to be affected when the actual word does not match the ink colour. On the other hand, when a high proportion of incongruent words are presented during a block of trials, the participant will actively avoid using the actual word to help the colour naming task. Since the subject is consciously trying to ignore the word, an incongruent word does not affect the subject as severely and the Stroop Effect is decreased.

To prevent yourself from getting confused, ask yourself whether you would be more or less surprised if an incongruent word was presented during the Stroop Task experiment. If there are more congruent words, an incongruent word would surprise you more, increase the time you require to perform the colour-naming task, and result in a bigger Stroop Effect.

Testing your Understanding

1) Which of the following lists of words in a Stroop Task experiment would result in the <u>GREATEST</u> Stroop effect?

a) 100% congruent words
b) 75% congruent words and 25% incongruent words
c) 50% congruent words and 50% incongruent words
d) 100% incongruent words

Answer: B. A higher percentage of congruent words will lead to an increased Stroop Effect. Participants will be using the actual word to help name the ink colour. Thus, when an incongruent word appears, participants will also use the word to name its ink color. However, since the word and its corresponding ink color do not match, their strategy will actually interfere with their ability to perform the task. Statement C is incorrect because a 50-50 chance of congruent and incongruent words will not encourage participants to adopt a particular strategy. Therefore, an incongruent word won't be especially surprising or cause the **greatest** Stroop Effect. A is not correct because the Stroop Effect only occurs when incongruent words are present. D is also incorrect because an entire array of incongruent words will cause participants to adopt a strategy to actively ignore the word itself, thus decreasing the Stroop Effect.

2) It has been noted that people suffering from attention-deficit hyperactivity disorder (ADHD) do significantly worse on Stroop tasks (i.e. are more affected by the Stroop Effect) than the average person. Why might this be the case?

Answer: The Stroop task requires a person to focus their attention on naming the color of the ink used to type a word while ignoring the word itself. However, it is the actual word which automatically captures an individual's attention. It has been hypothesized that people with ADHD do worse because they are either unable to ignore the interfering word and color information or they cannot direct attention and effort towards naming the ink color. In more general terms, people suffering from ADHD may be less able to direct conscious attention to tasks or may be less able to inhibit automatic attention processes.

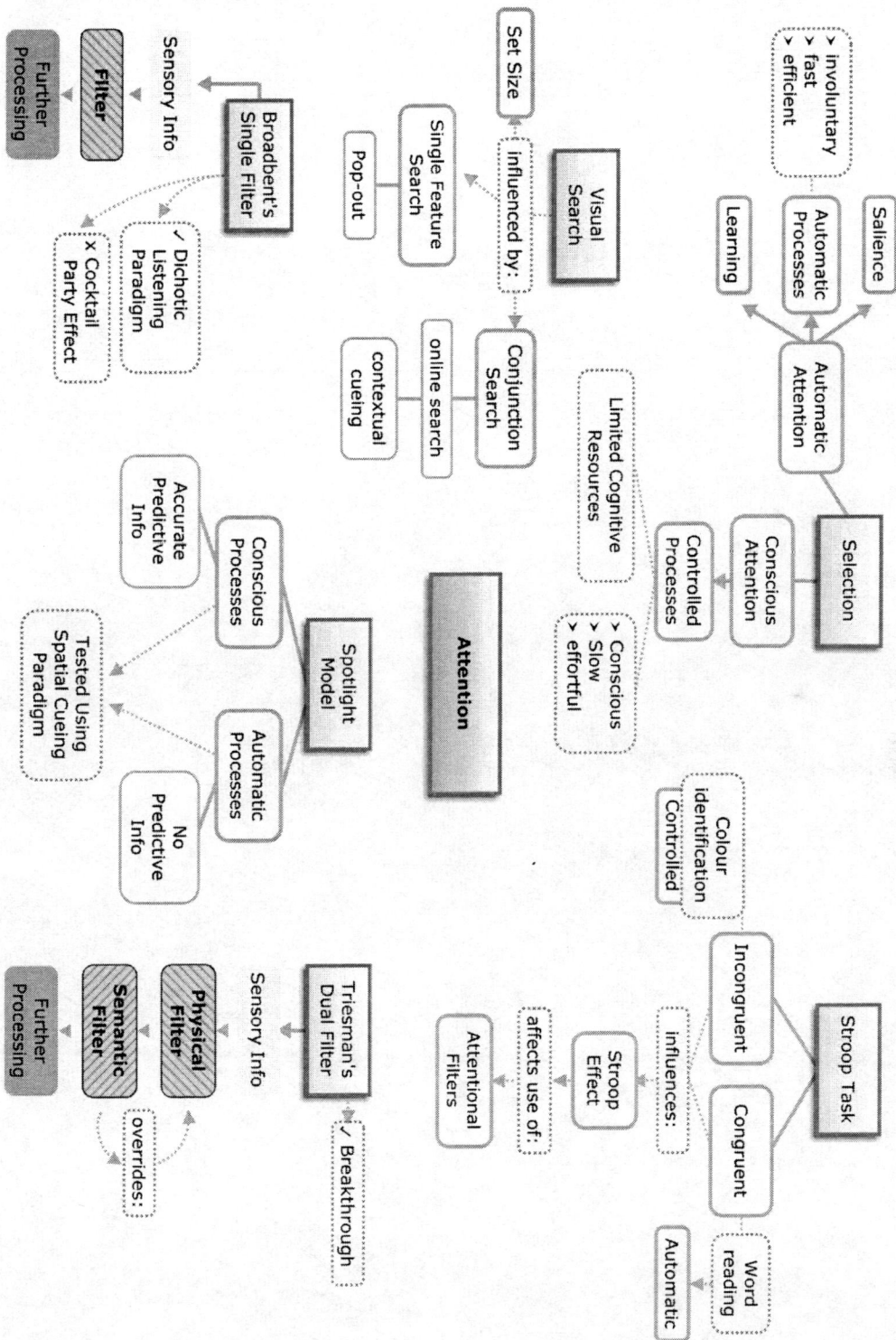

WEEK of Nov 10: MEMORY

"It's not only people who go to a therapist who might implant a false memory or those who witness an accident and whose memory can be distorted who can have a false memory. Memory is very vulnerable and malleable. People are not always aware of the choices they make."
- Elizabeth Loftus, prominent memory researcher

In 2002, police were investigating the case of the Washington Sniper, a serial killer targeting random victims in Washington, DC. Several eyewitnesses came forward claiming that a suspicious white van was spotted near the scene of several of the murders. However, no white van was found to be related to the murders. In fact, the murderer's vehicle was a dark green car. How did this memory of a white van come to be? Intriguing work from Dr. Elizabeth Loftus suggests that our memories are far from perfect and we are even susceptible to false memories. Loftus noted that an early eyewitness had participated in a media interview mentioning a white van. This idea stuck and subsequent eyewitnesses also made the same claim. Consider the following simple experiment run by Loftus. Subjects were told they were going to evaluate an advertisement, fill out several questionnaires and answer questions about a trip to Disneyland. The first group read a generic Disneyland ad that mentioned no cartoon characters. The second group read the same ad and was exposed to a 4-foot-tall cardboard figure of Bugs Bunny that was casually placed in the interview room. No mention was made of Bugs Bunny. The third, or Bugs group, read a fake Disneyland ad featuring Bugs Bunny. The fourth, or double exposure group, read the fake ad and also saw the cardboard rabbit. Thirty percent of subjects in the Bugs group later said they remembered or knew they had met and shook hands with Bugs Bunny when they visited Disneyland and 40 percent of the people in the double exposure group reported the same. The scenario described in the ad never occurred because Bugs Bunny is a Warner Brothers cartoon character and wouldn't be featured on Walt Disney property. Such research suggests that memory is far from an exact replica of reality; it is more accurate to say that memory is a reconstructive process. This has interesting implications for the reliability of eyewitness testimony and leads to more questions about the very nature of memory.

Braun, K. A., Ellis, R., & Loftus, E. F. (2002). Make my memory: How advertising can change our memories of the past. *Psychology & Marketing*, *19*(1), 1-23.

Weekly Checklist:
- ☐ **Web Module to watch: Memory**
- ☐ **Readings: Chapter 5 (sections 5-9)**
- ☐ **AVE Quiz 9**

Upper Year Courses:
If you enjoyed the content in this week's module, consider taking the following upper year courses:
- PSYCH 3FA3 The Neurobiology of Learning and Memory
- PSYCH 3II3 Cognitive Development
- PSYCH 3VV3 Human Memory

Module 9: Memory – Outline

Unit 1: Introduction to Memory

Common Memory Metaphors

Memory Metaphors

Metaphors allow psychologists to describe and understand memory.

Video camera: _____
_____
_____

Filing cabinet: _____
_____
_____

Computer: _____
_____
_____

Problems with Memory Metaphors

Data
- Stored data is identical to inputted information.
- Retrieved data is identical to inputted information.

Memory
- Stored memory includes personal details and interpretations
- Retrieved memory may be altered or lost

Assumptions of memory metaphors: _____
_____
_____
_____
_____
_____
_____
_____
_____
_____
_____

Unit 2: The Basics of Memory

The Importance of Cues

Hermann Ebbinghaus

Ebbinghaus was an influential researcher who made important findings in the areas of memory and forgetting.

_____
_____
_____
_____
_____
_____
_____
_____
_____
_____

Recall & Recognition

Recall Test

Recognition Test

VS.

Old? New?
- ☐☐ Shoes
- ☐☐ Pants
- ☐☐ Belt
- ☐☐ Socks
- ☐☐ Parakeet

- Both test ability to remember items from encoding phase.

Describe a basic memory task:

_____

_____

_____

_____

Describe the difference between recall and recognition:

_____

_____

_____

_____

Hermann Ebbinghaus

- zif
- tol
- dal
- puh
- riz
- lel

Number of Words Remembered

Time

Describe Ebbinghaus' research and findings:

_____

_____

_____

_____

_____

_____

_____

_____

_____

_____

_____

Unit 3: The Multi-Store Model

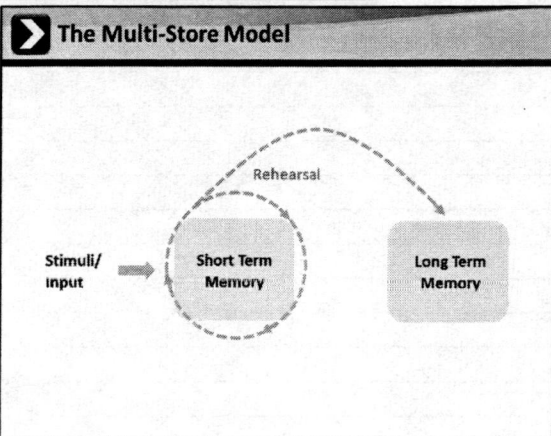

The Multi-Store Model

Rehearsal

Stimuli/ Input → Short Term Memory → Long Term Memory

Models allow researchers to understand difficult concepts such as memory.
Describe the multi-store model:

_____

_____

_____

_____

_____

_____

_____

_____

_____

_____

_____

_____

_____

_____

_____

_____

_____

_____

_____

_____

George Miller

Short-term Memory Capacity

$$7 \pm 2$$

→ 7-2
→ 7
→ 7+2

Short-term Memory

Capacity: _____

_____

_____

Rehearsal: _____

_____

_____

Chunking: _____

_____

_____

_____

Unit 4: The Serial Position Curve

Primacy

Primacy Effect

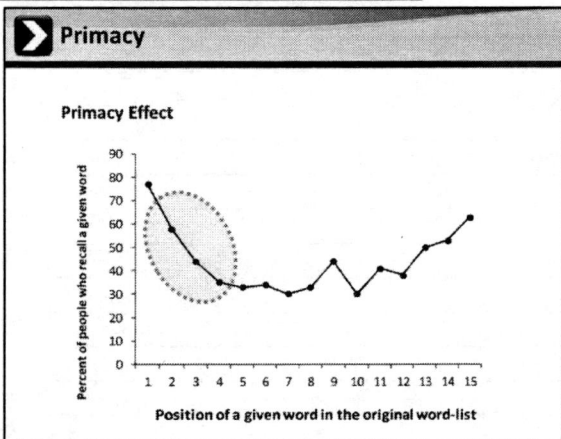

Percent of people who recall a given word

Position of a given word in the original word-list

The serial position curve describes a typical pattern of remembering when participants recall a list of memorized words.

Primacy effect: _____

_____

_____

Describe why the primacy effect occurs: ____

_____

_____

_____

_____

_____

Recency

Recency Effect

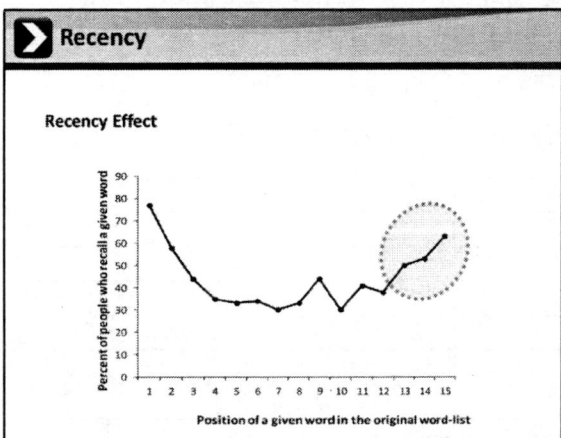

Percent of people who recall a given word

Position of a given word in the original word-list

Recency effect: _____

_____

_____

Describe why the recency effect occurs: ____

_____

_____

_____

_____

_____

_____

_____

_____

_____

_____

_____

_____

_____

_____

_____

_____

153

Improving Primacy

Increasing time between item presentations increases:

- Amount of times each item can be repeated
- Probability of item being stored in long term memory
- Performance recalling first couple of items

Diminishing Recency

Group 1
- Recall after performing different task for 30s

Group 2
- Recall after 30s silent interval

Group 3
- Immediately recall

- Recency effect diminished
- Recency effect
- Recency effect

Unit 5: The Levels of Processing Principle

Levels of Processing

Shallow level
- Encode physical characteristics
- Encoding requires little effort
- Poor memory performance

Deeper level
- Encode semantic characteristics
- Encoding requires significant effort
- Better memory performance

_____
_____
_____
_____
_____
_____

Describe how the primacy effect can be enhanced: _____
_____
_____
_____
_____
_____
_____
_____
_____
_____
_____

Describe how the recency effect can be diminished: _____
_____
_____
_____
_____
_____
_____
_____
_____
_____
_____

The Levels of Processing Model describes several stages at which memories can be encoded.

Describe the differences between shallow and deep levels of encoding: _____
_____
_____
_____
_____
_____
_____
_____
_____
_____
_____
_____
_____
_____

Levels of Processing

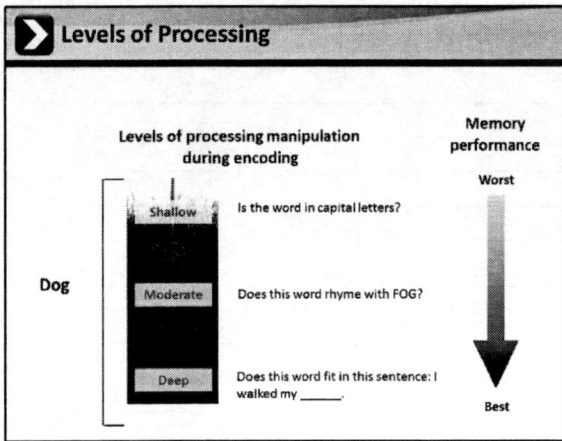

Levels of processing manipulation during encoding

Dog

Shallow	Is the word in capital letters?
Moderate	Does this word rhyme with FOG?
Deep	Does this word fit in this sentence: I walked my _____ .

Memory performance

Worst → Best

Craik and Lockhart: _____

The influence of levels of processing in daily life: _____

Unit 6: Encoding Specifity

Encoding Specificity

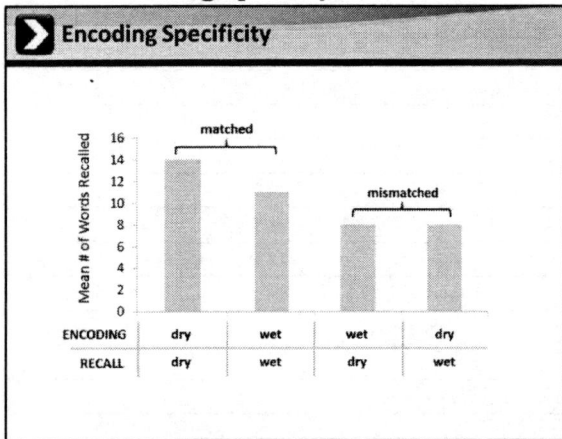

ENCODING	dry	wet	wet	dry
RECALL	dry	wet	dry	wet
	matched		mismatched	

Memories incorporate contextual information associated with specific experiences.
Encoding specificity: _____

Scuba divers experiment: _____

Unit 7: Memory Illusions and Fluency

Loftus and False Memories

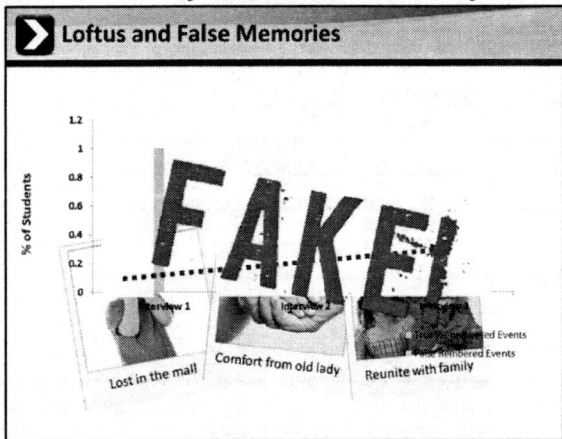

FAKE!

Lost in the mall — Comfort from old lady — Reunite with family

Our memories are prone to errors meaning that we do not store an exact record of past events.
Lost in a mall experiment: _____

Vending machine experiment: _____

Fluency

Definition

Fluency
- The ease with which an experience is processed, some experiences are easier (more fluent) than others

Attribution

Definition

Attribution
- Judgment tying together causes with effects

Becoming Famous Overnight

Phase I:
Pronunciation Task

- Robert Green
- Alana Cross
- Jeffery Lamb
- Lee Smith
- Deborah Man
- Katherine Smith
- etc.

Group A
No Delay

Group B
24 Hour Delay

Names from Pronunciation List

Novel Names

Phase II:
Fame Rating Task

- Julia Roberts
- Deborah Man
- Joe Kim
- Alana Cross
- Brian Wilson
- Kyle Penn
- etc.

Fluency: _____
_____
_____
_____
_____
_____
_____
_____
_____
_____
_____

Attribution: _____
_____
_____
_____
_____
_____
_____
_____
_____
_____
_____
_____

Becoming famous overnight experiment: ___
_____
_____
_____

Methods: _____
_____
_____
_____
_____
_____
_____
_____
_____
_____
_____
_____
_____
_____
_____

Becoming Famous Overnight

Phase I:
Pronunciation Task

- Robert Green
- Alana Cross
- Jeffery Lamb
- Lee Smith
- Deborah Man
- Katherine Smith
- etc.

Group A
No Delay

Group B
24 Hour Delay

- Names from Pronunciation List
- Novel Names

Phase II:
Fame Rating Task

- Julia Roberts
- Deborah Man
- Joe Kim
- Alana Cross
- Brian Wilson
- Kyle Penn
- etc.

Results:

_____

_____

_____

_____

_____

_____

_____

Implications:

_____

_____

_____

_____

_____

Unit 8: Conclusion

Problems with Memory Metaphors

Data

- Stored data is identical to inputted information.
- Retrieved data is identical to inputted information.

Memory

- Stored memory includes personal details and interpretations
- Retrieved memory may be altered or lost

Conclusion

Module 9: Memory – Courseware Exercise

Now that you've learned about the serial position curve, it's time for you to try it out for yourself. You will be running your very own memory experiment, replicating the findings of the serial position curve. It will not be a well-controlled experiment, but it is good for demonstrative purposes. (As a bonus, try to think of how you would correctly design this experiment if you were to run it in the lab).

Here is your task:
1. Using MS PowerPoint or paper, create a slideshow of 10 different words (each word on a different slide).
2. Recruit 3 to 5 friends for your experiment.
3. Have each participant simply sit and view the slideshow.
4. At the end of each slideshow, present them with a list of 20 words (10 will be the words presented in your slideshow, but not in the same order, and 10 will be novel words)
5. Have each participant circle the 10 words that they remember from the slideshow.
6. If you want a more accurate measure from each participant, repeat this procedure a few times per participant with new word slideshows.
7. Now, tally up each participant's results. Use the table below to do so. For instance: if the participant circled the word that was presented first, put an X under that serial position for that block of data.
8. Next, calculate the overall mean accuracy for each serial position across all participants.

Block	Accuracy by Serial Position									
	1	2	3	4	5	6	7	8	9	10
Example	X	X	X					X	X	
1										
2										
3										
4										
5										
6										
7										
8										
9										
10										
Mean										

(In the example, the participant correctly remembered the words presented 1st, 2nd, 3rd, 8th, and 9th)

Finally, plot your mean accuracy results in the graph below to see if they resemble the serial position curve.

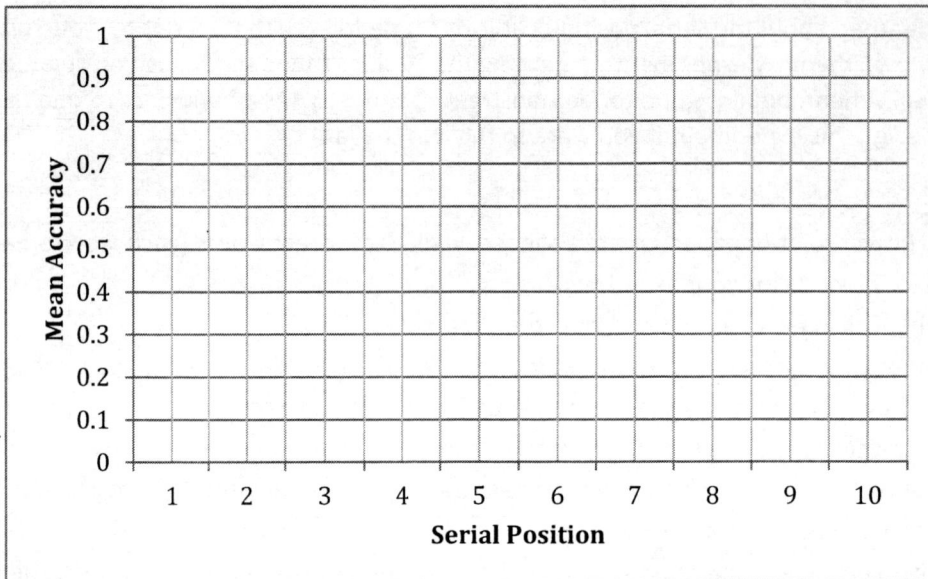

Do the results of your experiment resemble the serial position curve? If so, explain the main characteristics of the curve and why this serial position effect is occurring.

If your results differed from those shown in lecture for the serial position curve, explain how they differ and identify any factors in your experimental design that may have been affecting your results.

_____

_____

_____

_____

_____

_____

_____

Module 9: Memory – Review Questions

1) Describe Ebbinghaus' contribution to memory research.
   a. Draw the forgetting curve

2) Describe and draw the multi-store model of memory.

3) What is chunking? What is the benefit of chunking?

4) Read the following scenarios and determine how each affects primacy or recency in memory recall.

   a. You have been studying all day for a midterm, right up until the moment you leave for the testing room. On your way to the midterm, you call your friend to discuss plans for the evening.

   b. To attempt to learn all the students' names in tutorial, your TA leads an activity in which all students stand in a circle to play a simple game: The first person must say his or her own name. Then, the next person must say the first person's name and their own name. The next person must say the names of two people before and their own name. This continues around the circle until each student has gone and it's the TA's turn to recite every name. You are surprised to see that she is best at recalling the first few names!

   c. In an experiment, a researcher has subjects view a list of words. One group of subjects is allowed to view the list for 30 seconds while the other group is allowed to view it for 2 minutes.

   d. Your friend refuses to talk to anyone in the hall while waiting to write exams. Instead, he studies his notes right up until the exam begins.

5) What does encoding specificity suggest about memory? Describe some results suggesting the presence of encoding specificity in memory.

6) Describe Jacoby's experiment on "becoming famous overnight". What do the results suggest about memory encoding?

Module 9: Memory – Quiz Question

Which of the following would be most effective at manipulating the serial position curve?

- a. Increasing presentation time increases recall of all the words, so the curve becomes flat. (9%)
- b. Increasing the list size will diminish the primacy effect, but not the recency effect. (7%)
- c. Presentation of a distractor following the encoding phase will conserve the shape of the curve, but equally decrease the recall of all words. (14%)
- d. Presentation of a distractor following the encoding phase will decrease the recall of words presented at the end of the list. (70%)

This is a question used on the Memory and Attention Avenue Quiz from 2011 that evaluated students' understanding of the serial position curve. In order to succeed on this question, students should have reviewed what they know about the serial position curve and its important features such as the primacy and recency effects. Let's examine each of the options to determine where students may have gone wrong:

- a. This option reflects a common misunderstanding about the serial position curve. The serial position curve depicts memory performance based on that item's position in the list. Presenting words every 5 seconds instead of 2 seconds (increasing presentation time) will not flatten out the curve, it will increase recall only for items presented at the beginning of the list. Therefore, this option is incorrect.
- b. This option reflects a common misunderstanding about the serial position curve. The serial position curve depicts memory performance based on that item's position in the list. Increasing the list size, such as presenting 50 words instead of 20, will not diminish the primacy effect.
- c. This option requires a thorough understanding of why the primacy and recency effects occur. The primacy effect occurs because the first words on the list have more time for rehearsal, allowing them to move from short-term to long-term memory where they remain. The recency effect occurs because the last words on the list have not been replaced in short-term memory by new information. Information in short-term memory is often replaced since it can only hold 7±2 chunks of information. Once you understand these phenomena, it is clear to see that the presentation of a distractor following the encoding phase will *not* affect those words that have moved to long-term memory. As a result, this option is incorrect.
- d. Following the logic above (c), the presentation of a distractor following the encoding phase will *not* affect those words that have moved to long-term memory but *will* replace the most recently presented words from short term memory, thereby decreasing recall of words presented at the end of the list. As a result, this option is correct. (**Correct!**)

This leaves us with **option D** as the correct answer.

Key Terms

Attribution	False Memory	Recall Test
Chunking	Fluency	Recency Effect
Cue	Levels of Processing Model	Recognition Test
Encoding Specificity	Multi-store model	Serial Position Curve
False Fame Effect	Primacy Effect	

Module 9: Memory – Bottleneck Concepts

Primacy Effect vs. Recency Effect
Encoding Specificity
False Fame Effect
Attributional View of Memory

False Fame Effect

To understand this concept, we must first understand how fluency and attribution play a role in false memories. Fluency is the ease with which information is processed. Typically, familiar pieces of information are more easily processed. For example, it is easy to understand a scientific concept if you have encountered it before. Usually, when something is fluent, you can explain why it is more easily processed (e.g. The name, Dr. Joe Kim, will be fluent to you because he is your psychology professor). The process of explaining why a piece of information is processed with great fluency is known as attribution. You are linking the ease of processing to a cause. Sometimes, however, you cannot explain or remember why something or someone seems familiar or fluent. For example, why does that individual in a coffee shop look familiar to you? You can still make an attribution (e.g. that person might be in one of my classes, they are wearing the same shirt as one of my friends etc.), but that attribution may be incorrect and lead to a false memory.

Fluency and attribution can also be used to explain the false fame effect. Two groups of participants (A & B) read and pronounced a list of non-famous names. Group A was given a second list of names 24 hours later and were asked to give each name a fame rating. On this list were famous names, somewhat famous names, non-famous names from the list presented the day before, and novel non-famous names. Group B also received this second list; however, it was given to them immediately following the reading and pronunciation task. Interestingly, groups A and B rated the names similarly, except for one thing. Group A rated the non-famous names from the first list to be more famous than Group B. This is known as the false fame effect. This phenomenon can be explained through fluency and attribution. Since both groups had seen the non-famous names on the first list, the names would seem familiar and fluent to them. Participants from Group B, having just read the names, were able to make the correct attribution and say that the fluency occurred because they had just read the names. However, Group A, after a 24 hour delay, were more likely to forget where they had seen the name before. Thus, they were more likely to make an incorrect attribution and conclude that the name seemed familiar because it was somewhat famous. This effect can play out in many instances in your everyday life; take a look at the examples below and see if you can think of your own.

Test your Understanding:

1) Alicia was writing a multiple choice IntroPsych exam and noticed one of the response options contained the term "Encoding Attribution". This term seemed familiar but she could not remember where she had encountered it. Nevertheless, since she seemed to recognize it, she thought maybe that was the correct answer. She later found out that it was an incorrect response.

a) Explain why Alicia incorrectly selected Encoding Attribution as her answer. Use the terms fluency and attribution in your answer.

Answer: The phrase "Encoding Attribution" is similar to the terms "encoding specificity" and "attribution" found in this week's module. This is why the term "Encoding Attribution" was fluent to Alicia. Since Alicia did not remember the terms "encoding specificity" and "attribution" from the module, she was not able to correctly attribute the fluency of this novel term. Instead, she incorrectly thought the term had been covered in lecture and chose it as her answer.

b) Has this incidence happened to you before? What are some techniques to prevent this instance of incorrect attribution from occurring?

2) Paul introduces you to his friend Ryan whom you have never met before. In which of these scenarios would you be most likely to say that you've met Ryan before?

a) You saw a photo of Ryan on your friend's phone a week before meeting Ryan.
b) Ryan looks like a friend you just bumped into at the supermarket.
c) Another friend showed you a picture of Ryan last night.
d) You have never seen a photo of Ryan before.

The correct answer is A. In this scenario, Ryan would seem fluent and familiar. Since you saw the picture a week ago, which was greater length of time compared to the other scenarios, you would most likely have forgotten where you saw Ryan and incorrectly attribute the fluency to having met Ryan before.

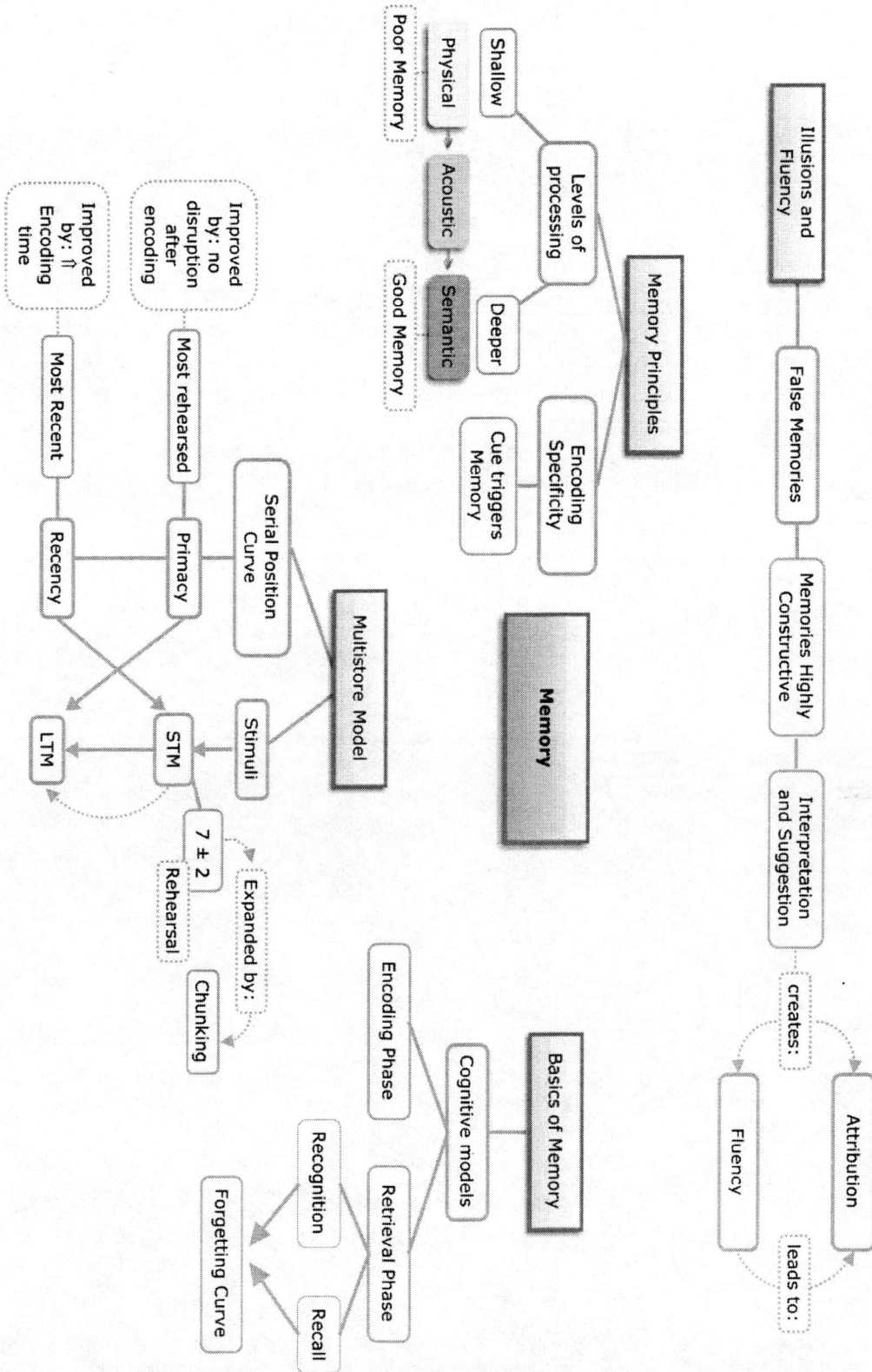

Memory

Memory Principles

Levels of processing
- Shallow → Physical → Poor Memory
- Deeper → Acoustic → Semantic → Good Memory

Encoding Specificity
- Cue triggers Memory

Multistore Model

Serial Position Curve
- Most Recent → Recency
- Most rehearsed → Primacy

- Improved by: ⇑ Encoding time → Most Recent
- Improved by: no disruption after encoding → Most rehearsed

- Stimuli → STM → LTM
- STM: 7 ± 2 → Expanded by: Chunking
- Rehearsal

Basics of Memory

Cognitive models
- Encoding Phase
- Retrieval Phase
  - Recognition → Forgetting Curve
  - Recall → Forgetting Curve

Illusions and Fluency

- False Memories → Memories Highly Constructive → Interpretation and Suggestion
- creates: Attribution
- Fluency
- leads to:

WEEK of Nov 17: FORMING IMPRESSIONS

"The psychological opposition of irreconcilable ideas, held simultaneously by one individual, created a motivating force that would lead, under proper conditions, to the adjustment of beliefs to fit prior behaviour instead of changing behaviour to express beliefs – the sequence conventionally assumed."
-Leon Festinger on cognitive dissonance

Group membership is often steeped in ritual. Although hazing is now banned on most university campuses, new university students are still referred to as "frosh" until they attend their first class. Having passed this threshold, the new members join the larger university community. Although such a tradition may be fairly innocuous, the rites of passage for membership in other groups can involve more severe rituals. After completing 10 training jumps, US Marine paratroopers receive their golden wings, a metal badge with two half-inch pins. During the blood pinning ceremony, the pin is thumped into a marine's chest. A controversy was sparked when amateur videotapes aired on news reports showing the ritual in full. The tapes depicted dozens of marines taking turns punching, pounding and grinding the gold pins into the bloody chests of new initiates who scream in pain. One video plainly showed a senior officer watching as his men are being abused. Although blood pinning has officially ended, rites of passage are not likely to be completely eliminated. Hazing rituals are viewed by senior group members as an integral part of the tradition. Even initiates who have gone through the ritual insist it is an important part of bonding. A practical solution may be to shift the focus to regulation. As summarized by General John Shalikashvili (Chairman of the Joint Chiefs of Staff at the time), "We are always going to try to fix human behaviour, but we are not always going to have 100% success."

Gleick, E. (1997, Feb. 10). Marine blood sports. *Time, 149*(6), 30.

Festinger, L., & Carlsmith, J. M. (1959). Cognitive consequences of forced compliance. *The Journal of Abnormal and Social Psychology, 58*(2), 203.

Weekly Checklist:
- ☐ **Web Module to watch: Forming Impressions**
- ☐ **Readings: Chapter 6 (sections 1-2)**
- ☐ **AVE Quiz 10**

Upper Year Courses:
If you enjoyed the content in this week's module, consider taking the following upper year courses:
- PSYCH 2C03 Social Psychology
- PSYCH 3CR3 Attitudes and Persuasion

Module 10: Forming Impressions – Outline

Unit 2: Attribution Theories

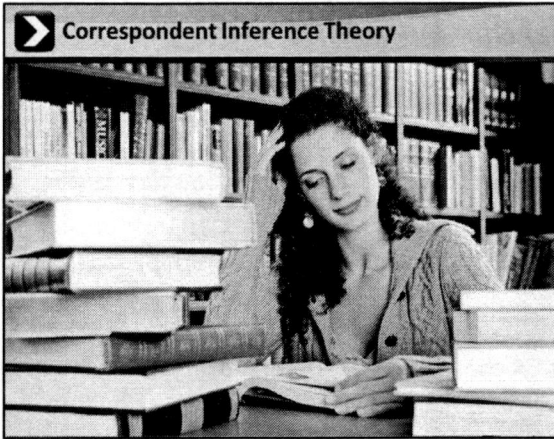

Correspondent Inference Theory

The Correspondent Inference Theory suggests that we attribute others' behaviours based on three variables.

Degree of choice: _____
_____
_____
_____

Expectation: _____
_____
_____

Intended consequences: _____
_____
_____

Covariation Theory

The Covariation Theory determines if we will attribute someone's behaviour to their disposition or their situation.

Consistency: _____
_____

Distinctiveness: _____
_____
_____
_____

Consensus: _____
_____
_____
_____

Unit 3: The Fundamental Attribution Error

The Fundamental Attribution Error

Definition

Fundamental Attribution Error
■ Tendency to over-value dispositional factors for the observed behaviours of others while under-valuing situational factors

The Fundamental Attribution Error describes our tendency to attribute others' behaviours to dispositional factors.

_____
_____
_____
_____
_____
_____
_____
_____
_____
_____

The Actor/Observer Effect

Cultural Differences

American adults tended to attribute behaviours to personal factors more often than situational

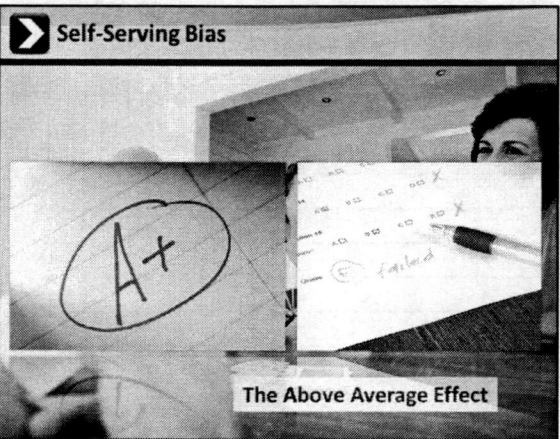

Self-Serving Bias

A+

failed

The Above Average Effect

_____
_____
_____
_____
_____
_____
_____

Being aware of situational pressures that influence our behaviours, we attribute our actions to our situation.

Actor/Observer Effect: _____
_____
_____
_____

Supporting evidence from teen drivers: _____
_____
_____
_____
_____
_____

Cultural differences in FAE for young children: ___
_____
_____
_____

Cultural differences in FAE for teens and adults: __
_____
_____
_____

Attributions made by Olympic athletes from different cultures: _____
_____
_____
_____

Self-Serving Bias allows us to perceive ourselves favourably in numerous situations.

Above Average Effect: _____
_____
_____

Attributions made for positive events: _____
_____
_____

Attributions made for negative events: _____
_____
_____
_____
_____
_____
_____
_____
_____

168

Unit 4: Cognitive Heuristics

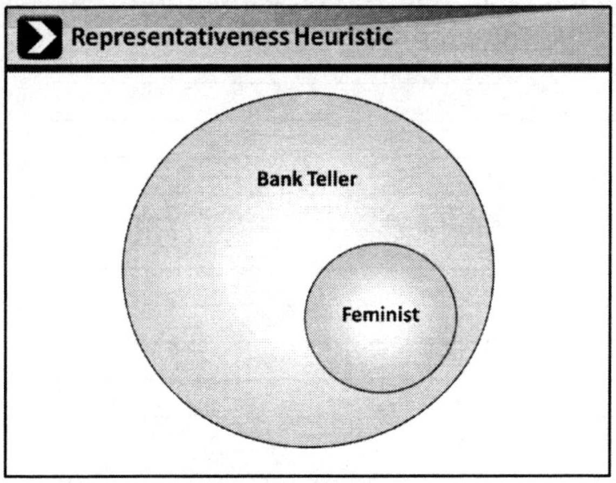

Representativeness Heuristic

Bank Teller

Feminist

Along with helping us make quick decisions, heuristics are also used make social judgements.

Representativeness Heuristic: _____

_____
_____
_____

The bank teller example: _____

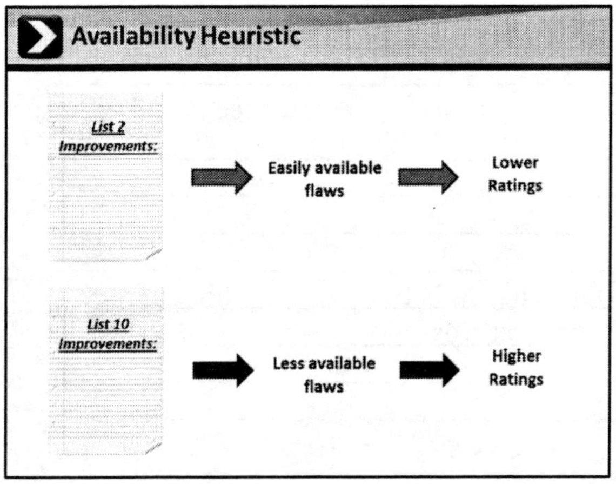

_____
_____
_____
_____
_____
_____

Availability Heuristic

List 2 Improvements: → Easily available flaws → Lower Ratings

List 10 Improvements: → Less available flaws → Higher Ratings

Availability Heuristic: _____

_____
_____
_____

The course evaluation example: _____

_____
_____
_____
_____
_____
_____
_____

Unit 5: Relationships

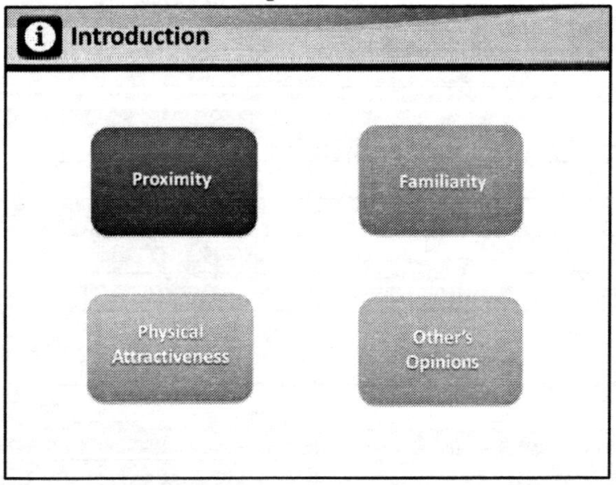

Introduction

Proximity

Familiarity

Physical Attractiveness

Other's Opinions

A combination of four factors can predict the degree of attraction between two individuals.

_____
_____
_____
_____
_____
_____
_____
_____
_____
_____
_____
_____
_____
_____

_____
_____
_____
_____
_____

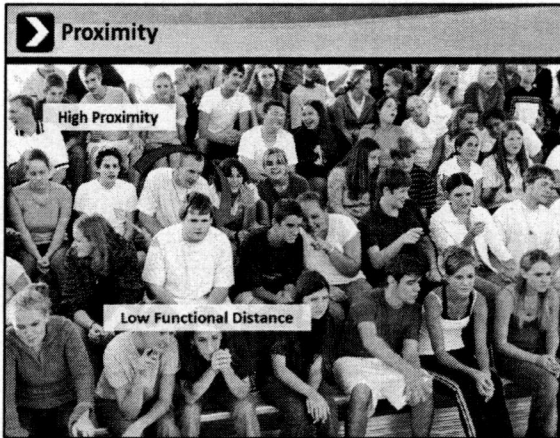

Proximity

High Proximity

Low Functional Distance

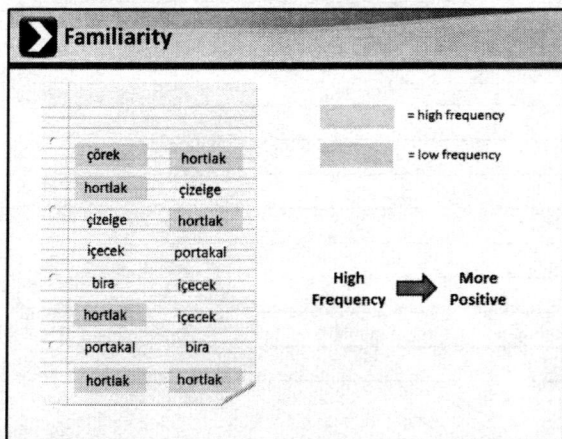

Familiarity

çörek	hortlak
hortlak	çizeige
çizeige	hortlak
içecek	portakal
bira	içecek
hortlak	içecek
portakal	bira
hortlak	hortlak

= high frequency

= low frequency

High Frequency → More Positive

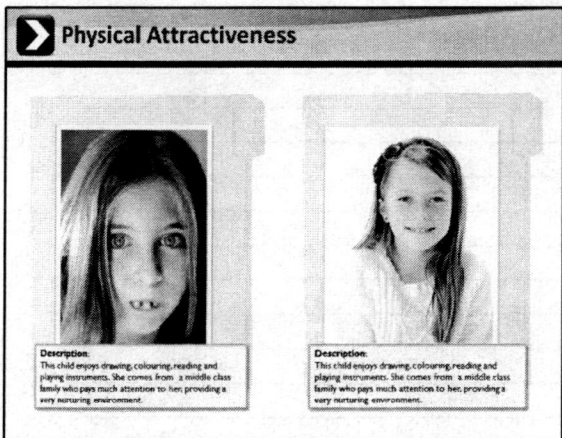

Physical Attractiveness

Description:
This child enjoys drawing, colouring, reading and playing instruments. She comes from a middle class family who pays much attention to her, providing a very nurturing environment.

Description:
This child enjoys drawing, colouring, reading and playing instruments. She comes from a middle class family who pays much attention to her, providing a very nurturing environment.

_____
_____
_____
_____
_____
_____
_____

Proximity: _____
_____
_____

Functional distance: _____
_____

How does anticipating an interaction with someone influence their attraction ratings?
_____
_____
_____
_____
_____

The Mere Exposure Effect: _____
_____

Describe the foreign words experiment: _____
_____
_____
_____

Why do we rate a mirror image of ourselves as being more attractive than a correct image?
_____
_____
_____
_____

Physical attractiveness: _____
_____
_____
_____
_____

How does an individual's attractiveness impact their ratings in other domains?
_____
_____
_____
_____
_____
_____
_____
_____
_____
_____
_____
_____
_____
_____
_____

Liking Those Who Like Us

Describe the effect of self-esteem on being attracted to those who find us attractive:

_____
_____
_____
_____
_____
_____
_____
_____

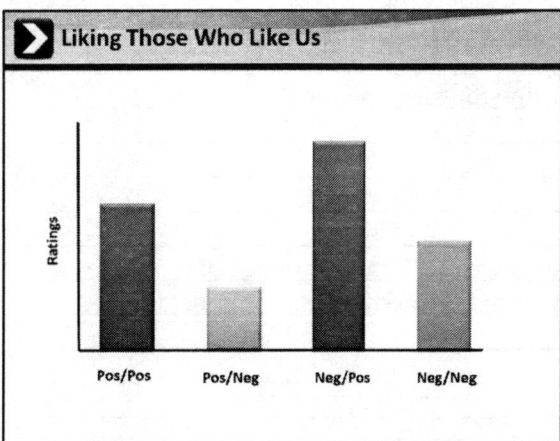

Liking Those Who Like Us

Ratings

Pos/Pos   Pos/Neg   Neg/Pos   Neg/Neg

How do others' opinions of us influence our opinions of them?

_____
_____
_____
_____
_____
_____
_____
_____
_____

Unit 6: Conclusion

Conclusion

Conclusion

_____
_____
_____
_____
_____
_____
_____
_____
_____
_____
_____
_____
_____
_____
_____

_____
_____
_____
_____
_____
_____

Module 10: Forming Impressions – Courseware Exercise

It is Sarah's first day at McMaster University and she is desperate to make a good impression in order to make new friends. Using the principles you've learned on forming impressions, describe four strategies that Sarah can use to present herself favourably to a new classmate.

1) Principle: _____

   Strategy:_____

   _____

2) Principle: _____

   Strategy:_____

   _____

3) Principle: _____

   Strategy:_____

   _____

4) Principle: _____

   Strategy:_____

   _____

In using these strategies, Sarah has become friendly with two girls at her school. Monica is in Sarah's psychology class, a subject that Sarah enjoys and excels at. Sarah knows Maria from her physics class, a mandatory course with an awful professor, a course she'd rather not be taking. Recently, Sarah won 2 tickets to a concert from a local radio station. In choosing which friend to bring, Sarah decides on Monica because she'll be more fun to go to the concert with.

What heuristic is Sarah making use of and why is it possible that her assumption is incorrect?

_____

_____

_____

_____

Module 10: Forming Impressions – Review Questions

1) According to correspondent inference theory, which three factors influences our analysis of others' behaviour?

2) Using covariation theory, determine if the male student's behaviour below is situational or dispositional:

   Yesterday, as you walked into your Biology lecture a few minutes late, a male student was standing at the front of the lecture hall bragging to the whole class about all his accomplishments and avid devotion to community involvement. When you take your seat beside your friend, she tells you that the her friend speaking at the front is a really great, humble guy who would be the best candidate for president of the student's union.

3) .a. What is the fundamental attribution error?

   b. Answer the following question regarding FAE:

   Cheryl, a Canadian bank-teller, is out for lunch with her 9 year-old son Matthew and her new co-worker Meili who recently immigrated from China. While waiting to order, they overhear their waitress speaking rudely to her other customers. What are each person's likely responses to the behaviour of the waitress?

   a. **Cheryl and Matthew** are most likely to assume that the waitress has had a bad day; **Meili** is most likely to assume that the waitress is just a rude person.
   b. **Cheryl** is equally likely to assume that the waitress has had a bad day or that she is just a rude person; **Matthew** is most likely to assume that the waitress has had a bad day; **Meili** is most likely to assume the waitress is just a rude person.
   c. **Cheryl** is most likely to assume the waitress has had a bad day; **Matthew and Meili** are most likely to assume the waitress is just a rude person.
   d. **Cheryl** is most likely to assume that the waitress is just a rude person; **Matthew** is equally likely to assume that the waitress has had a bad day or that she is just a rude person; **Meili** is most likely to assume the waitress has had a bad day.

   b. What accounts for cross-cultural differences in FAE?

4) How do the availability and representativeness heuristics apply to forming impressions of others?

5) List four factors that play an important in our liking of others. Describe each factor.

Module 10: Forming Impressions – Quiz Question

Katherine has been watching a lot of television to find examples of correspondent inference theory. As she channel surfs, she witnesses many different behaviours. In which case has Katherine correctly analyzed behaviour according to the correspondent inference theory?

a. President Obama is supporting the legalization of marijuana and Katherine thinks that he is so passionate that his position on the matter must be his own. (39%)
b. Homer Simpson is wearing glasses and acting very intelligent, and Katherine thinks that his attire and behaviours lack consistency with Homer's usual self. (21%)
c. Walter White is yelling at one of his friends in his house, at work and at a restaurant, which Katherine feels represents his behaviour being present across a variety of situations. (13%)
d. Dexter Morgan finds a man tied down on a table and, instead of helping him, Dexter hits him on the head. Katherine feels that others would not do the same thing in such a situation. (26%)

This is a question from the Psych 1X03 Forming Impressions Avenue Quiz taken by students in 2011. In this question, students are asked to apply the three aspects of **correspondent inference theory** and to differentiate these from the three variables used in **covariation theory**. In correspondent inference theory, the variables used to analyze behaviour are *degree of choice*, *expectation*, and *intended consequences*. In covariation theory, the variables are *consistency*, *distinctiveness*, and *consensus*. Now, let's look at each option in turn:

a. This option suggests that Katherine believes President Obama's position in this debate is his own. This reflects the variable *degree of choice* in **correspondent inference theory**, because she believes that he was not forced to argue for the legalization of marijuana and chose this position. Degree of choice *is* a variable in correspondent inference theory and thus A is the correct answer. (**Correct!**)
b. This option applies the variable of *consistency* in **covariation theory**, because Katherine thinks that Homer does not often behave in this manner in his daily life. Since the question requires us to consider correspondent inference theory, this option is incorrect.
c. This option describes *distinctiveness* in **covariation theory**, since by asking the question "Does Walter yell at his friends in a variety of situations or just in this situation?" Katherine can determine that this behaviour is not distinct for any one situation. However, the question requires us to consider correspondent inference theory which makes this option incorrect.
d. Here, Katherine analyzing Dexter's behaviour using the *consensus* variable of **covariation theory** by asking "Would others behaviour similarly in this situation?" Since the question requires us to consider correspondent inference theory, this option is incorrect.

We can see that only **option A** uses an aspect of correspondent inference theory to analyze behaviour. The other three options use variables from covariation theory, which is not what the question has asked for! Thus **A** is the only correct option.

Key Terms

Above Average Effect	Correspondent Inference Theory	Fundamental Attribution Error
Actor/Observer Effect	Covariation Theory	Other's Opinions
Attribution Theory	Cultural Differences	Physical Attractiveness
Consensus	Distinctiveness	Proximity
Consistency	Familiarity	Self Serving Bias

Module 10: Forming Impressions – Bottleneck Concept

Fundamental Attribution Error
Availability Heuristic vs. Representativeness Heuristic
Correspondent Inference Theory vs. Covariation Theory
Mere Exposure Effect

Availability Heuristic vs. Representativeness Heuristic

Both of these heuristics may seem similar at first, but there are clear differences between them. To start, a heuristic is a decision-making shortcut we use to help us make quick decisions or impressions about elements in our surroundings. The availability heuristic, when used in the context of forming impressions, represents how we often use the most readily available information when forming an opinion about others. For example, if your group of friends aren't fond of a particular class and regularly complain about its difficulty, you will immediately think of these negative opinions when asked about the class. You would probably develop an impression that the class is terrible and difficult based of the information most readily available to you.

The representativeness heuristic, on the other hand, describes our assumptions of people or things based on an internal representation that we have of certain groups. Stereotypes are a form of the representativeness heuristic. For example, when you see a well dressed student on campus, you might assume that they belong to the school of business because in your mind, all business students dress well. When trying to differentiate between the two, ask yourself, "Is the person using the most readily available information about a topic/person to make their judgment? Or are they making conclusions based on an internal representation of that topic/person?"

Test your Understanding:

1) Is the following situation an example of the availability or representativeness heuristic? Why?

a) Philip was Karly's online TA for the physiology course Karly took in first year. A year later, Karly met Philip at the Charity Ball and found out that he was her TA in first year. Meeting Philip for the first time in person, Karly was surprised to see that he was outgoing and looked like an average guy. She imagined him to have glasses, be reserved, and not come out to social events because he was a TA.

Answer: This is an example of the representativeness heuristic. Karly made assumptions about Philip because she knew he was a TA and she possessed an internal representation of what TAs look like and how they act.

b) A new restaurant opened up down the street from Dave's house. Many of Dave's neighbours told him that the food was exceptional. Dave recommended the restaurant to his friend, Kyle, who wanted to find a good place to eat. The food ended up being mediocre at best. Dave later found out that his neighbours like everything they eat.

Answer: This is an example of the availability heuristic. Dave used the most available information to him to form an impression of restaurant. He heard the positive comments from his neighbours and just assumed that the restaurant had good food.

2) Explain how the media can promote stereotypes, with reference to the availability and representativeness heuristics.

Answer: For some people, the media may be the primary source of information about things or people. This means that whatever is shown in the media may be the most readily available information for consumers. Through the availability heuristic, people may incorrectly form an impression of certain things and even certain groups leading to stereotypes of those around us. The media may also promote representations of certain groups that are internalized by viewers and lead to the development or reinforcement of stereotypes.

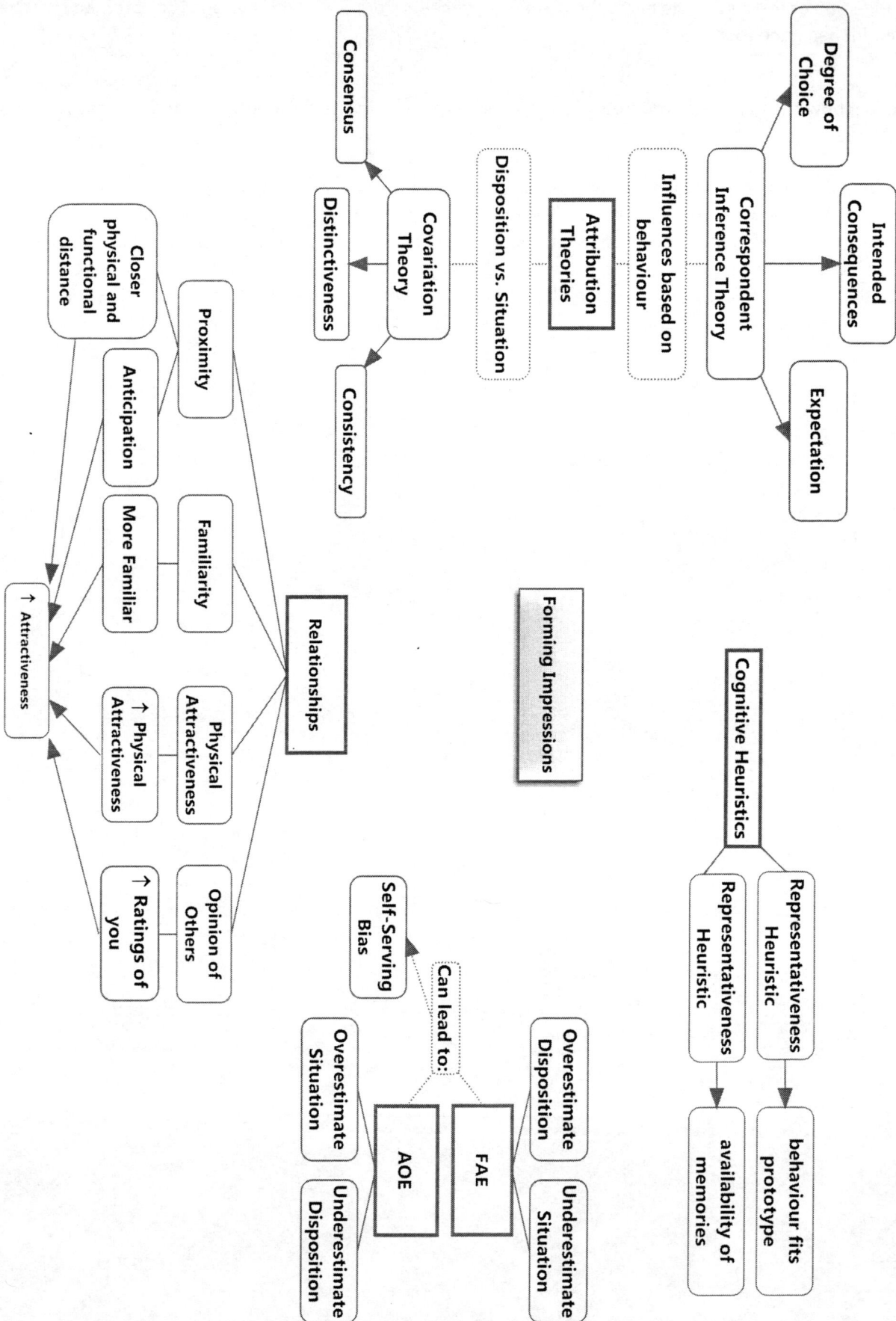

WEEK of Nov 24: INFLUENCE OF OTHERS

"I was not surprised that it happened. I have exact, parallel pictures of prisoners with bags over their heads."
- Dr. Philip Zimbardo, comparing the Abu Ghraib prison conditions to his infamous Stanford Prison study.

In this week's web modules, we will explore some fascinating experiments in which subjects under controlled settings could be made to act in strange and sometimes sadistic ways. We like to think that our own behaviour (guided by dispositional factors) would prevent us from acting as many subjects did in these experiments. Are such findings applicable to real world settings? Outside the lab, can ordinary people be made to act in bizarre ways at the command of another person? The scene was a busy Kentucky Fried Chicken restaurant in an affluent Manchester, New Hampshire neighbourhood. A young supervisor was overseeing his energetic staff through the lunchtime rush when he received a call directly from corporate headquarters. This was unusual in itself, but the message from corporate executive Jeff Anderson was even more unusual. Jeff calmly explained that a toxic chemical had been released from the overhead sprinkler system. He ordered the supervisor to gather all the employees. Over the speaker phone, Jeff explained that everyone must strip down and urinate on each other to neutralize the chemical. To encourage the employees, at one point Jeff said, "I need you to be strong, I need you to be brave and do exactly what I say." Police arrived in the parking lot to witness a strange scene of several naked women, panicking and soaked in each other's urine. In fact, there was no toxic chemical and the employees were never in any danger. Jeff Anderson turned out to be a prankster named "Dex" who had perpetrated several such deceptions and posted them on the internet. Authorities in several US cities are considering whether Dex should be charged and extradited to face trial.

Verma, S. (2009, Aug. 8). Criminal or comedy? Online anonymity lets cruel pranksters take jokes too far. *Globe and Mail*, 1-2.

Haney, C, Banks, W.C., & Zimbardo, P.G. (1973). Interpersonal dynamics in a simulated prison. *International Journal of Criminology and Penology, 1* , 69-97.

Weekly Checklist:
- ☐ **Web Modules to watch: Influence of Others 1 and 2**
- ☐ **Readings: Chapter 6 (sections 3-7)**
- ☐ **AVE Quiz 11**

Upper Year Courses:
If you enjoyed the content in this week's module, consider taking the following upper year courses:
- PSYCH 2C03 Social Psychology
- PSYCH 3AC3 Human Sexuality
- PSYCH 3CD3 Intergroup Relations

Module 11: Influence of Others I - Outline

Unit 2: Presence of Others

Norman Triplett

Definition

Co-Actor
- Another individual performing the same task

Audience
- A group of people watching an individual perform a task

Social Facilitation
- The increased performance that occurs in the presence of co-actors or an audience

Zajonc's Resolution

- Presence of others increases arousal to improve performance on simple tasks, and decrease performance on complex tasks

Unit 3: Social Learning Theory

_____
_____
_____
_____
_____
_____
_____
_____
_____
_____
_____
_____
_____
_____

Norman Triplett conducted systematic studies to demonstrate that people behave differently in the presence of others.

Triplett's findings: _____
_____
_____
_____
_____
_____

Complications of Triplett's hypothesis: _____
_____
_____
_____

Describe Zajonc's resolution of conflicting data from social facilitation studies: _____
_____
_____
_____
_____
_____
_____
_____
_____
_____
_____

Albert Bandura, a proponent of Social Learning Theory, suggested that we learn our behaviours from those around us.

Describe how Social Learning Theory can be differentiated from conditioning: _____
_____

What were the main findings from the Bobo Doll experiment and its follow up? _____
_____
_____
_____
_____
_____
_____
_____

Unit 4: Conformity

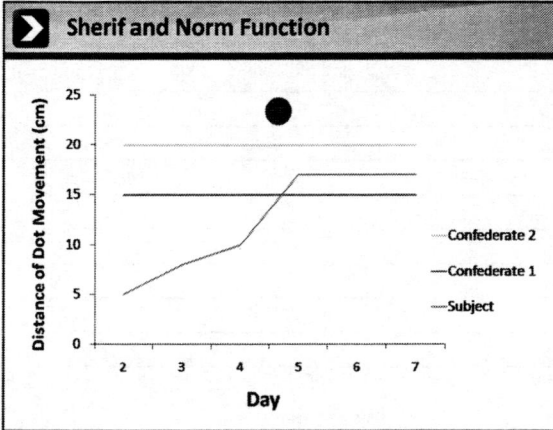

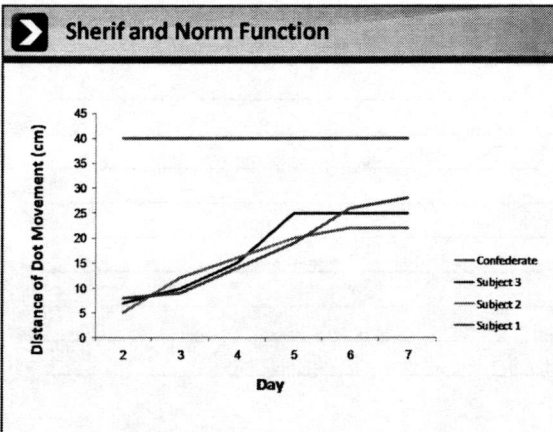

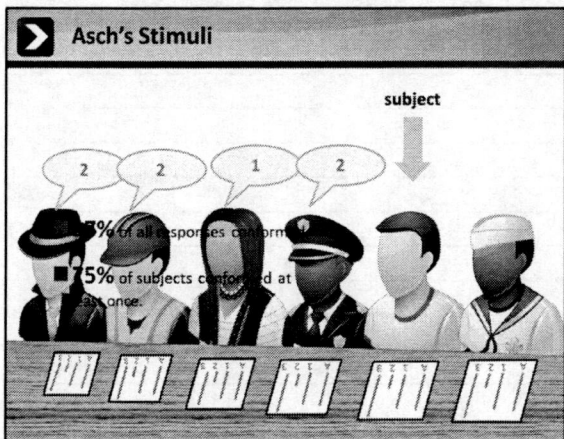

Muzafer Sherif studied norm formation and conformity using the autokinetic effect.

Describe the autokinetic effect:

_____

_____

_____

Describe Sherif's experimental setup:

_____

_____

_____

_____

_____

_____

_____

What were the results of Sherif's experiments?

_____

_____

_____

_____

_____

_____

_____

_____

_____

_____

_____

_____

_____

Solomon Asch furthered Sherif's research by determining why individuals conform to group pressures.

Describe Asch's experimental setup:

_____

_____

_____

What were the results of Asch's experiments?

_____

_____

_____

Describe the difference between normative and comparative functions:

_____

_____

_____

_____

_____

_____

_____

180

Unit 5: Group Dynamics

The Risky Shift

Of these options, what do you think is the lowest probability of success you would consider acceptable for Helen to write the novel?

1-10%	10-20%	20-30%	40-50%	50-60%	60-70%	70-80%

The Risky Shift

(Bar chart: Probability of Success vs. Decision Level — Individual ≈ 0.4, Group ≈ 0.3)

Group Polarization

(Line chart: Level of Risk vs. Individual/Group — Helen line increasing from mid to high; Roger line decreasing from mid to low)

_____

_____

_____

_____

_____

_____

_____

James Stoner systematically studied group decision making to determine how group decisions differ from those made individually.

Describe Stoner's experimental setup:

_____

_____

_____

_____

_____

_____

_____

_____

_____

Describe the Risky Shift and the types of situations where it is a concern:

_____

_____

_____

_____

_____

_____

_____

_____

_____

_____

_____

How does the concept of Group Polarization provide a better explanation of group decision making when compared to the Risky Shift?

_____

_____

_____

_____

_____

_____

_____

_____

_____

_____

_____

_____

_____

_____

_____

Groupthink

Definition

Groupthink
- A group decision making environment that occurs when group cohesiveness becomes so strong it overrides realistic appraisals of reality and alternative opinions

Groupthink

- Be impartial
- Critical evaluation
  - Devil's advocate
- Subdivide the group
- Provide a second chance

Unit 6: The Bystander Effect

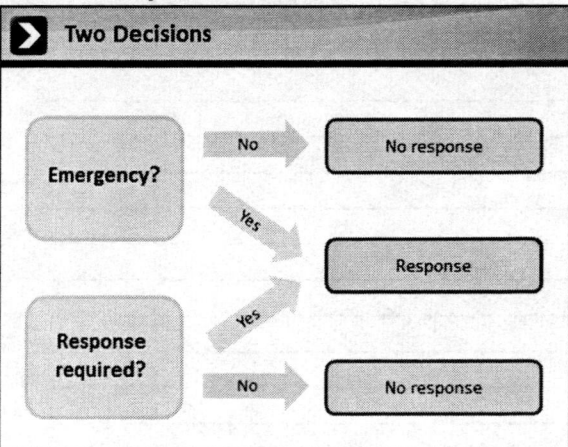

Two Decisions

Emergency? → No → No response

Emergency? → Yes → Response

Response required? → Yes → Response

Response required? → No → No response

_____
_____
_____
_____
_____

Describe Groupthink and its implications:
_____
_____
_____
_____
_____
_____
_____
_____
_____
_____
_____

What are some ways in which Groupthink can be prevented?
_____
_____
_____
_____
_____
_____
_____
_____
_____
_____

Describe an individual's thought process when deciding whether to respond in an emergency:
_____
_____
_____
_____
_____
_____
_____
_____
_____
_____
_____
_____
_____
_____
_____

Two Decisions

Definition

Collective Ignorance
- When each individual in a group see nobody responding in a given situation, they conclude that the situation is not an emergency

Describe supporting evidence for the concept of Collective Ignorance and the decision-making stage at which it influences bystanders:

_____
_____
_____
_____
_____
_____
_____
_____
_____

Two Decisions

Definition

Diffusion of Responsibility
- In deciding whether we have to act, we determine that someone else in the group is more qualified.

Describe supporting evidence for the concept of Diffusion of Responsiblity and its influence on bystanders:

_____
_____
_____
_____
_____

How can individuals ensure they receive help when needed?

_____
_____
_____

Social Loafing

Describe how the concept of Social Loafing is related to Diffusion of Responsibility and provide evidence supporting this phenomenon:

_____
_____
_____
_____
_____
_____
_____
_____
_____
_____
_____
_____
_____

_____
_____
_____
_____
_____
_____

Module 11: Influence of Others I – Courseware Exercise

On January 28th 1986, the Space Shuttle Challenger broke apart just after lift-off, resulting in disintegration of the shuttle and the deaths of all crew members on board. Though the disaster was due to a faulty O-ring seal, further investigation afterwards revealed that NASA's decision-making process had been seriously flawed. The project managers had known about the faulty design 9 years before the launch, but failed to correctly address the problem. Furthermore, NASA engineers had warned their managers that launching on a cold morning, as they had, could be disastrous for the space shuttle. However, these managers failed to act on and report this information to their superiors.

The Rogers Commission Report was created in order to investigate the incident and offer suggestions to NASA for future space missions. You are encouraged to do a bit of extra research on the Challenger disaster and the Rogers Commission recommendations.

Knowing that a large group of people were working on this one project and having learned about the dynamics of group work, what do you think was at the root of NASA's problematic decision-making process?

_____

_____

_____

Plan out a course of action for decision-making that NASA could have taken, incorporating all of the solutions to groupthink that were presented in lecture.

_____

_____

_____

_____

_____

Module 11: Influence of Others I – Review Questions

1) What did Triplett observe regarding bicycle racers?
    a. Why is Triplett's conclusion not enough?
    b. Describe Zajonc's contribution.

2) Why is Bandura's Bobo Doll experiment important?

3) Describe Sherif's study of Norm Formation.

4) Describe Solomon Asch's famous experiment on conformity.

5) Describe Deutsch and Gerard's modification to Asch's experiment.

6) Outline the difference between the normative and comparative functions of a group.

7) What is group polarization? How is it different from the risky shift?

8) What do Darley and Latane's results suggest about helping behaviour in groups?
    a. What is diffusion of responsibility and how can it be broken?

9) What is social loafing?

Module 11: Influence of Others I – Quiz Question

Which of the following is a major difference between collective ignorance and diffusion of responsibility?

a) Collective ignorance is a form of group decision-making that results in a strong opinion without taking into account alternate options, whereas diffusion of responsibility is a lack of response while in a group of individuals to an emergency. (9%)

b) Collective ignorance occurs when analyzing a situation, whereas diffusion of responsibility occurs when choosing whether to respond to a situation. (69%)

c) Collective ignorance involves concluding a situation is not an emergency due to the absence of emergency personnel present, whereas diffusion of responsibility involves not responding to a situation. (18%)

d) Collective ignorance involves ignoring all situations that do not have personal involvement, whereas diffusion of responsibility is placing the blame on others when you failed to respond. (4%)

This is a question from the 2011 Psychology 1X03 Influence of Others Avenue Quiz. This question is testing students understanding of collective ignorance and diffusion of responsibility. In order to correctly answer this question, it is important to understand that collective ignorance occurs when an individual is unsure if the situation is an emergency and looks to others for an indication of the appropriate course of action. Diffusion of the responsibility, on the other hand, occurs when an individual has determined that the situation is an emergency but is unsure whether they should get involved or wait for another person to help. If we look at each option in turn, we will see that only one of these options fits that description:

A. This is clearly not describing collective ignorance as we just defined it; in fact, it is more closely describing the risky shift and other aspects of group dynamics. Option A does, however, correctly describe diffusion of responsibility.

B. This option appears to fit our description of these phenomena but, as is often the case in IntroPsych tests it does not use the exact definition, forcing students to draw their own conclusions and apply the concepts. (**Correct!**)

C. This option is a very tempting choice and misleads 18% of students, but there is one aspect of this option that makes it incorrect. When collective ignorance occurs, people in the situation do not base their judgement of whether it is an emergency on the presence of emergency personnel.

D. This is also an obvious misrepresentation of these two phenomena based on our previous definition. It is incorrect.

We can see after examining all the possible choices that **option B** best outlines the differences between collective ignorance and diffusion of responsibility.

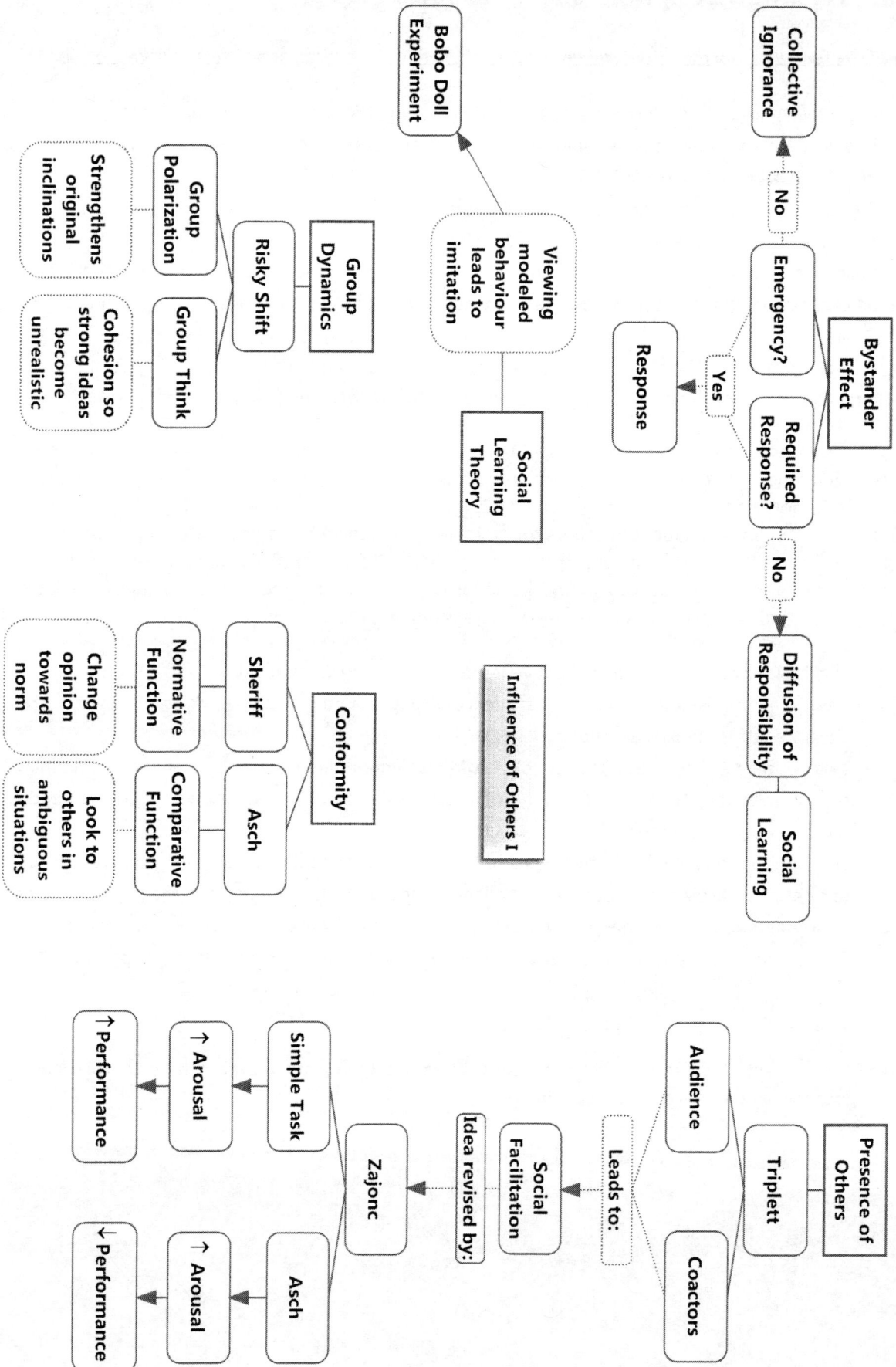

Influence of Others I

Bystander Effect

Emergency?
- No → Collective Ignorance
- Yes → Required Response?
  - Yes → Response
  - No → Diffusion of Responsibility → Social Learning

Social Learning Theory

Viewing modeled behaviour leads to imitation → Bobo Doll Experiment

Group Dynamics

Risky Shift
- Group Polarization → Strengthens original inclinations
- Group Think → Cohesion so strong ideas become unrealistic

Conformity

- Normative Function → Sheriff → Change opinion towards norm
- Comparative Function → Asch → Look to others in ambiguous situations

Presence of Others

Triplett
- Audience
- Coactors

Leads to: Social Facilitation

Idea revised by: Zajonc

- Simple Task → → Arousal → → Performance
- Asch → → Arousal → ← Performance

187

Module 11: Influence of Others II – Outline

Unit 2: Obedience

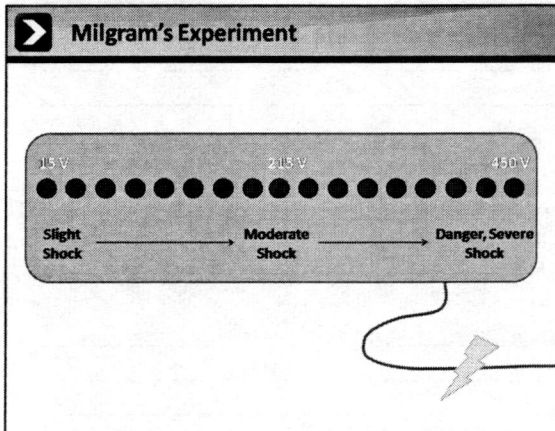

> **Milgram's Experiment**

jl5 V   2l5 V   450 V

Slight Shock   →   Moderate Shock   →   Danger, Severe Shock

> **Important Lessons**

Obedience (%) — 0, 20, 40, 60, 80

"Would you obey?"   Actual Obedience

> **Manipulations**

Obedience (%) — 0, 20, 40, 60, 80

In Person   Over the Phone

Stanley Milgram's experiment elucidated the effects of authority on human behaviour.

Milgram's experimental setup:

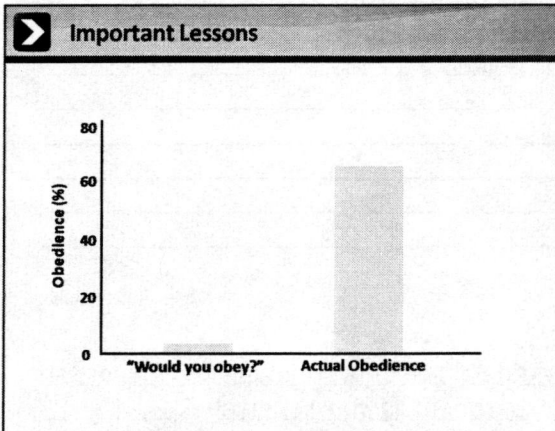

_____
_____
_____
_____
_____
_____
_____
_____
_____
_____

Results from Milgram's original shock experiment:

_____
_____
_____

Implications of Milgram's experiment:

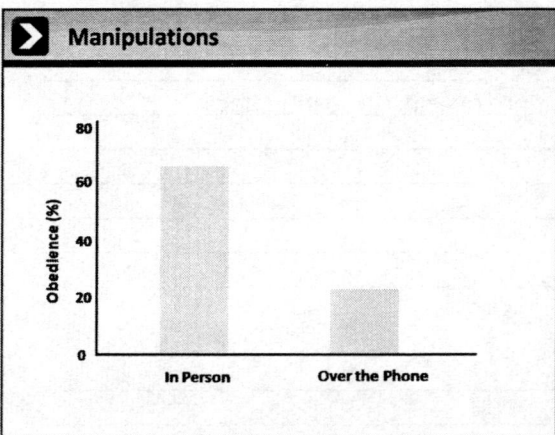

_____
_____
_____
_____
_____
_____
_____

How did Milgram change his original design and how did these manipulations impact the results?

_____
_____
_____
_____
_____
_____
_____
_____
_____
_____

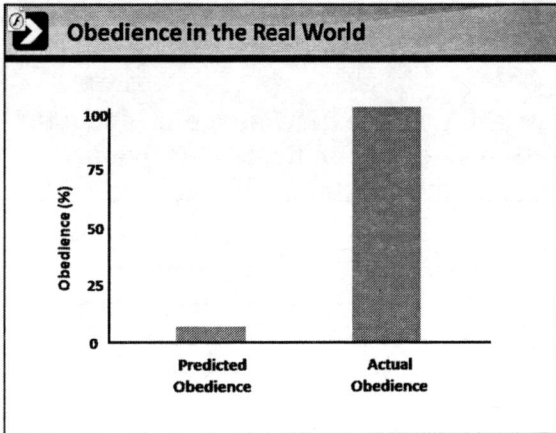
Obedience in the Real World

Unit 3: Cognitive Dissonance

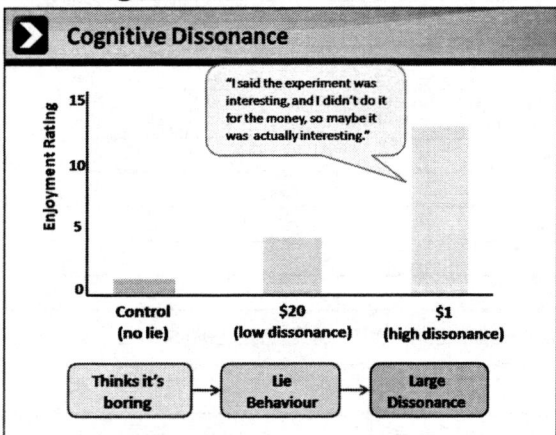
Cognitive Dissonance

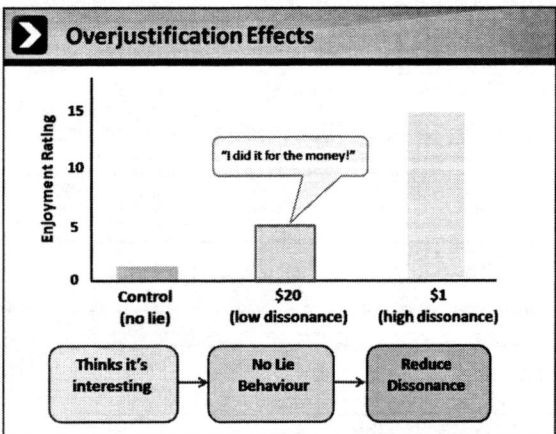
Overjustification Effects

_____
_____
_____
_____
_____
_____
_____
_____
_____

Describe the study by Holfing et al. and how it demonstrates the power of authority in a real world setting:
_____
_____
_____
_____
_____
_____
_____
_____
_____

What is cognitive dissonance?
_____
_____
_____
_____

How did the experiment conducted by Festinger and Carlsmith produce cognitive dissonance in participants?
_____
_____
_____
_____
_____
_____

Why did participants receiving less money rate the experiment as more enjoyable?
_____
_____
_____
_____
_____

What are overjustification effects?
_____
_____
_____
_____
_____
_____
_____

_____
_____
_____
_____
_____
_____

189

Unit 4: The Stanford Prison Experiment

Deinviduation

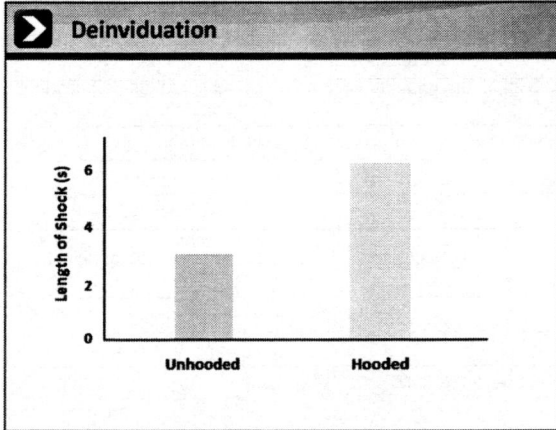

Bar graph: y-axis "Length of shock (s)" from 0 to 6; x-axis categories "Unhooded" and "Hooded"

Unit 5: Persuasion

The Communicator

Similarity OR Credibility?

Lifestyle Choices ➔ Similarity

Objective Fact ➔ Credibility

The Message

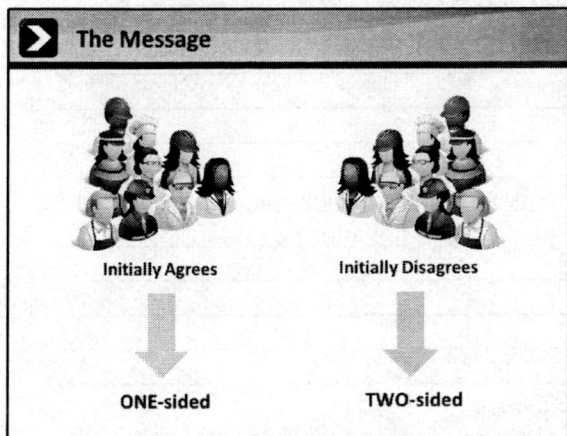

Initially Agrees ⬇ ONE-sided

Initially Disagrees ⬇ TWO-sided

_____
_____
_____
_____
_____

Describe the Stanford Prison Experiment:
_____
_____
_____
_____

In what manner did deindividuation contribute to the results?
_____
_____
_____
_____
_____

Several factors play in role in determining whether an individual can persuade others to behave in a certain manner.

Describe specific traits that may enhance a communicator's persuasiveness:
_____
_____
_____
_____
_____
_____
_____
_____

How can an audience's inclinations influence the contents of a persuasive message?
_____
_____
_____
_____
_____
_____
_____
_____
_____
_____
_____
_____
_____

The Audience

Definition

Central Appeal
- Well reasoned, factual, two-sided arguments
- Effective for intelligent audiences

Peripheral Appeal
- Well presented, easy to understand messages
- Effective for unintelligent audiences

How do the characteristics of the audience impact the contents of the message?

_____
_____
_____
_____
_____
_____
_____
_____
_____

Unit 6: Techniques in Persuasion

The Foot in the Door Effect

Definition

Foot in the Door Effect
- A gradual escalation of demands increases obedience

Describe the Foot in the Door Effect and evidence supporting this phenomenon:

_____
_____
_____
_____
_____
_____
_____
_____
_____
_____
_____

The Low-Ball Technique

Definition

Low Ball Technique
- An escalation of the terms of an agreement after someone has already agreed

Describe the Low-Ball Technique and evidence supporting this phenomenon:

_____
_____
_____
_____

What are some differences between the Foot in the Door Technique and the Low-Ball?

_____
_____
_____
_____

_____
_____
_____
_____
_____
_____

_____
_____
_____
_____
_____
_____

Module 11: Influence of Others II – Courseware Exercise

You've now learned about persuasion and how people can be convinced by others. This area of psychology has many practical applications, from conflict management to motivational speaking and from politics to marketing.

Imagine that you have been hired to design an advertising campaign for a small business that produces Psychology textbooks for university students. They want to become well-known and increase sales and have thus given you complete creative freedom to do so. You first decide to put together a commercial for the company. Using the principles that you learned in lecture, detail your plans for the commercial.

Who is your communicator? What should they look like? Should they be credible or similar? Do they have any important personality traits?

_____

_____

Consider your message. Should it be one-sided or two-sided? What type of appeal should be used and why?

_____

_____

Who is your audience? Are you appealing to a specific subpopulation of university students? What salient characteristics of your audience will be important to keep in mind?

_____

_____

Following the first phase of the advertising campaign, the company has done a bit of market research. Unfortunately this small company must charge a 10% more than a larger rival company. Research showed that 80% of students who bought textbooks from the small company felt some anxiety afterwards at having spent so much when they knew they were on a tight budget. Afraid of losing these customers to the larger company during textbook season next term, the company enlists your help for the second phase of the advertising campaign.

How might you use cognitive dissonance to your advantage and address this issue in your campaign? (hint: you don't want these students to change their behaviour, so you must influence an attitude change)

_____

_____

_____

Module 11: Influence of Others II – Review Questions

1) Describe Stanley Milgram's obedience experiment setup.
   a. What were the results of this experiment?
   b. What are some possible factors contributing to Milgram's result?
   c. What are the implications of this experiment?

2) In Festinger and Carlsmith's cognitive dissonance experiment, why did the $1 group rate the experiment more highly than the $20 group? Explain using cognitive dissonance and overjustification.

3) Describe the setup of the Stanford Prison Experiment.
   a. What were the results of this prison experiment?
   b. What are the implications of this experiment?

4) What makes a persuasive communicator?

5) Is a two-sided message better than one-sided?

6) What is the difference between central and peripheral appeal? When should either be used?

7) Describe the "Foot in the Door" technique. How is the foot in the door related to the low ball technique? Use your own example to illustrate the differences between the two.

Module 11: Influence of Others II – Quiz Question

Jeffrey is an artist who has created an unusual t-shirt design that may not appeal to everyone. While it does not resemble other popular designs, he would still like to get feedback to use to promote further sales. According to **cognitive dissonance theory**, which of the following marketing schemes would lead to Jeffrey receiving the most positive feedback?

a) Charging $20 per t-shirt with Jeffrey's design, allowing him to make a profit. (37%)
b) Charging $5 per t-shirt with Jeffrey's design, informing customers it's to cover the cost of the t-shirt only. (22%)
c) Giving the first 10 t-shirts away for free in hopes of better reviews. (15%)
d) Paying the first ten individuals $5 to take a t-shirt and provide feedback. (27%)

This question is from the 2011 Influence of Others I & II Avenue Quiz. It is a question that many students struggled with, so let us examine what the question is asking more closely. Jeffrey is trying to use cognitive dissonance theory to gain *positive* feedback for his t-shirt design. Cognitive dissonance occurs when one's behaviours do not match their attitude and can be anxiety-producing.

Looking at our options, we can see:

A. In this case, the customers must pay $20 for the t-shirt. The behaviour of paying that amount of money indicates that they must like the shirt; therefore, it would create cognitive dissonance if they provided negative feedback for the design—"Why would I pay $20 for the shirt if I didn't like it? It must be amazing!" Therefore, individuals in this situation are most likely to provide positive feedback because of cognitive dissonance. (**Correct!**)

B. In this option, the customer providing feedback is forced to pay for the shirt, but it is a small amount and only to cover the cost of the shirt. It is possible, in this case, that the person does not like the shirt very much, but because it is inexpensive and only allowing Jeffrey to break even. Mediocre or negative feedback (attitude) won't be at odds with paying $5 (behaviour).

C. In this option, accepting a free shirt does not mean the person likes the t-shirt design at all. They have not gone out of their way to obtain the t-shirt. Therefore, producing mediocre or negative feedback won't create any cognitive dissonance with their behaviour of accepting a free shirt.

D. This option involves the audience being paid to accept a t-shirt and provide feedback, so they again did not go out of their way to obtain the shirt. Again, there will be no cognitive dissonance between accepting money and providing negative feedback.

As a result, **option A** is the correct answer.

Key Terms

Asch's Stimuli	Comparative Function	Normative Function
Audience	Deindividuation	Overjustification Effects
Bobo Doll Experiment	Foot in the Door Effect	Peripheral Appeal
Bystander Effect	Group Polarization	Persuasion
Central Appeal	Groupthink	Risky Shift
Co-Actor	Low-Ball Effect	Social Loafing
Cognitive Dissonance	Message	Stanford Prison Experiment
Communicator	Milgram Experiment	Zajonc's Resolution

Module 11: Influence of Others I & II – Bottleneck Concepts

Normative vs. Comparative Function
Cognitive Dissonance
Foot in the Door Effect vs. Low-Ball Effect
The Bystander Effect

Normative vs. Comparative Function

Both normative and comparative functions describe ways in which others can influence our behaviour in a group setting. The normative function comes into play when, due to a fear of rejection, we conform to what others are doing. This may lead us to perform actions and behave in ways that are contrary to our beliefs. The influence of normative function can be evident in fashion trends, peer pressure to consume alcohol and drugs, and listening to certain types of popular music.

The comparative function, on the other hand, happens in ambiguous situations where we do not know what to do. To determine an appropriate response, we often observe others' behaviours and act accordingly. This can be seen when students receive an essay assignment without an explicit page limit. To determine approximately how many pages they should write, students will often resort to asking their peers. In this case, the peers provide information in an ambiguous situation. Importantly, students in this scenario are not conforming due to a fear of rejection. Other examples of the comparative function include: comparing your answers with a group of friends when completing a practice midterm, copying others' dance moves when you are uncertain of your own, and following a crowd during a fire drill because you do not know the location of the fire exit.

To differentiate between the normative and comparative functions, determine whether the subject is following others due to a fear of rejection or because he does not know what to do in that situation.

Test you Understanding

1) Identify if the following situations are an example of the normative or comparative functions.

a) Students at Creekside High School were asked to vote anonymously between two bands to play at their graduation. Since Karly did not know either band, she asked her friends which band they would vote for and ended up voting for the same band.

Answer: There was no fear of rejection because the vote was anonymous and no one who know who Karly voted for. She ended up voting for the same band as her friends because she did not know anything about either of the two bands. Karly's friends provided her with information in an ambiguous situation. This is an example of the comparative function.

b) Trevor has a severe fear of heights preventing him from going on roller coasters at amusement parks. One day, Trevor visited the park with all his friends who were especially excited to ride the roller coaster. Trevor, not wanting to look wimpy and afraid, decided to give the roller coaster a shot.

Answer: Trevor went on the roller coaster due to a fear of rejection from his friends. This is an example of the normative function.

2) Explain why in Deutsch and Gerard's (1955) manipulation of Asch's stimuli experiment, the researchers demonstrated the comparative function as opposed to the normative function.

Answer: Deutsch and Gerard had participants seated in their own cubicles, allowing them to provide anonymous answers. Participants also saw lights indicating the confederates' answers before they inputted their own response. Quite often, the subjects would go along with the confederates' incorrect answers, especially when the correct answer was ambiguous. Since participants' answers were anonymous, fear of rejection was not a significant factor. Therefore, the normative function did not come into play. Instead, this scenario is an example of the comparative function. The situation in Deutsch and Gerard's experiment created an ambiguous situation where two answers seemed possible leading subjects to refer to others' responses to determine the correct response.

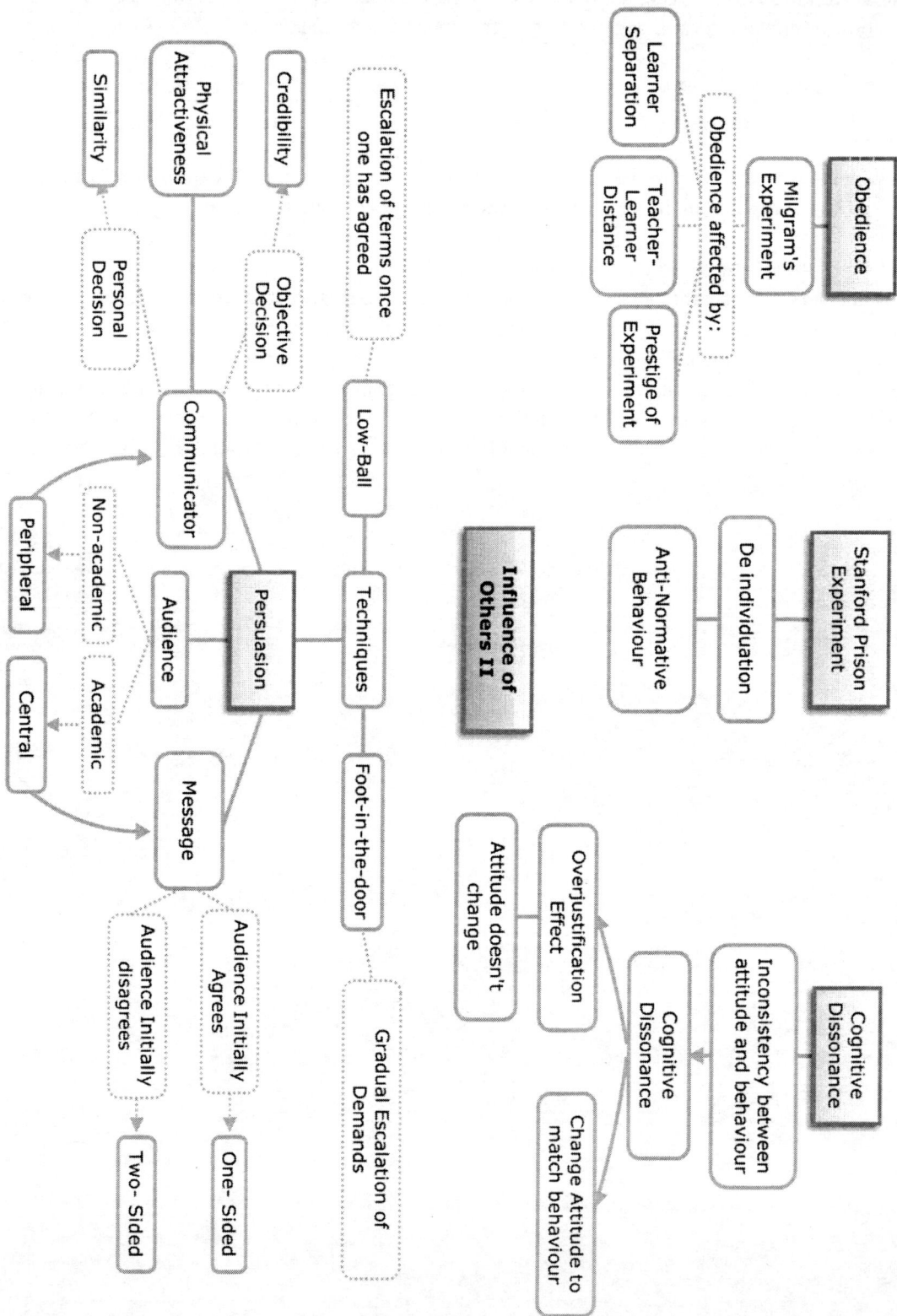

Influence of Others II

Obedience

Milgram's Experiment

Obedience affected by:
- Learner Separation
- Teacher-Learner Distance
- Prestige of Experiment

Stanford Prison Experiment

De individuation → Anti-Normative Behaviour

Cognitive Dissonance

Inconsistency between attitude and behaviour → Cognitive Dissonance
- Overjustification Effect → Attitude doesn't change
- Change Attitude to match behaviour

Persuasion

Communicator
- Credibility ← Objective Decision
- Physical Attractiveness
- Similarity ← Personal Decision

Audience
- Non-academic → Peripheral
- Academic → Central

Message
- Audience Initially disagrees → Two-Sided
- Audience Initially Agrees → One-Sided

Techniques
- Low-Ball — Escalation of terms once one has agreed
- Foot-in-the-door — Gradual Escalation of Demands

ANSWER KEY

ACTIVITIES AND PRACTICE QUESTIONS

Module 1: Levels of Analysis – Courseware Exercise

1. Individual answers may vary. Please consult your TA if you require assistance.

Module 1: Levels of Analysis - Review Questions

1. Explain human behaviour by focussing on the physiological mechanisms of the brain.
   a. Neuroimaging techniques include X-rays, CT scans and structural and functional MRIs. Structural neuroimaging techniques provide information about the physical make-up of the brain. Functional neuroimaging techniques provide information about what the brain is actually doing.
2. Learning researchers aim to examine the effects of an environment on behaviour.
   a. According to the behaviourism perspective, overt behaviour is the only valid means of measure in psychology. Behaviourism is unique because behaviourists view the mind as a black box; what happens within is off limits for scientific study.
3. Ethics are an important concern for social psychologists as they often work with human subjects. When creating artificial social situations, social psychologists must consider the effects of their manipulations on both the mental and physical well being of their participants. In addition, in order to ensure that the artificial situations they create are controlled, social psychologists sometimes use the technique of deception. This means that they cannot inform participants about the true purpose of the experiment.
4.

	Similarities	Differences
Developmental Approach	-Studies the influence of genetic and environmental factors -Interested in changes in behaviour	-Looks at changes in behaviour across a lifespan -Explores how humans change and develop between conception and death
Evolutionary Approach	-Studies the influence of genetic and environmental factors -Interested in changes in behaviour	-Looks at changes in behaviour across many generations -Explores ultimate causation (how did this trait affect survival?)

5. a-iv; b-iii; c-i; d-v; e-ii.

Module 2: Research Methods I – Courseware Exercise

Lighting in Factories 1: 1. No control group. 2. No hypothesis. 3. Possibility of practice effect (since all subjects are put into changing conditions from week to week). 4. Potential for subject biases (since AMC has volunteered to participate in this study). 5. Small sample size.

Lighting in Factories 2: 1. Has a testable hypothesis. 2. Uses random sampling when selecting subjects. 3. Has a control group. 4. Uses a between-subjects design (to avoid the practice effect). 5. Has a large sample size.

1. **Problem**: The experimenter knows which factories will receive which condition. As a result, experimenter bias might occur.
   Solution: Use a double blind study, in which neither the experimenter nor the participants know which group each participant belong to.
2. **Problem**: The factories participating in this study are located in different countries. As a result, it is possible that the control groups may differ from the experimental groups in more ways than the independent variable alone.
   Solution: Have both a control and an experimental group present in each factory throughout the course of the experiment.

Module 2: Research Methods I – Review Questions

1. Please consult your TA if you require assistance with this activity.
2. A theory is a *general* set of ideas about the way the world works while a hypothesis makes *specific* predictions about a phenomenon. A hypothesis is a testable statement that is *guided* by theories.
3. a) **IV**: Breakfast consumption. **DV**: The amount of material learned in lecture.
   b) Between-subjects design. The independent variable is manipulated across two groups of participants. Participants only take part in one condition (they either eat breakfast or they don't). To make this into a within-subjects design, you could have the same group of students eat breakfast before watching the module one week and skip breakfast before watching the following week's module.
   c) Random sampling is important for a quality experiment as the sample of participants must accurately reflect the population you are studying. Choosing a sample at random from the entire population reduces the chance that results might be biased towards a specific group.
4. a
5. d

Module 2: Research Methods II – Courseware Exercise

1. The results from Professor A's class have a larger spread (and thus a larger standard deviation) than the results from Professor B's. This means that in Professor A's class some students scored very well, while others scored very poorly. In Professor B's class, for which the results have a smaller

198

spread, most students obtained very similar scores, despite the average being slightly lower than class A's. This suggests that all students learn similar amounts from Professor B. In contrast, only some students learn a great amount from Professor A, while others learn much less and perform very poorly. These results suggest that professor A's teaching methods only appeal to some but not all of his students.

2. Using inferential statistics, a t-test can be performed to in order to see if the difference in scores between Professor A's class and Professor B's class is statistically significant. If results are "not statistically significant" they are considered to be obtained by chance. If results of the t-test are "statistically significant", it would suggest that there is a real difference in scores between the two classes, which *may* suggest differences in teaching abilities of Professor A and Professor B. Keep in mind, many other factors could contribute to this difference between the classes apart from teaching ability!

Module 2: Research Methods II – Review Questions

1. Please refer to the "Using Histograms" section in module 3.
2. **Mean**: 5.6, **Median**: 6, **Mode**: 6
3.

	Advantages	Disadvantages
Mean	-Useful for statistical purposes -Most common measure of central tendency	-Influenced by outliers -Exact value may not appear in the data set
Median	-Not influenced by outliers	-Exact value may not appear in the data set
Mode	-Not influenced by outliers -Exact number appears in the data set -Can be used with non-numerical data sets	-Not representative of all data

4. d has the largest standard deviation of 20
   b has the smallest standard deviation of 3
5. While a population is the full group of individuals that you are seeking to study, a sample is a subset of that population of interest. An experimenter would prefer to conduct an experiment on a sample as conducting an experiment on an entire population (e.g. all undergraduate students) would be too costly and time-consuming and likely impossible. Inferential statistics allow us to use results from samples to make inferences about the overall, underlying populations.
6. Statistically significant results suggest that the difference between two or more groups is due to some true difference between them and not random variation.
7. Correlation is a measure of the strength of the relationship between two variables. We must be careful when

measuring correlations, as "correlation does not necessarily equal causation". Although variables may be strongly correlated, we cannot definitively state that the relationship between them is causing the observed effect.

Module 3: Classical Conditioning – Courseware Exercise

1. Answers may vary. The idea is that you would pair your product with something positive. For example, in the TV commercial, you could pair the presentation of the product with a popular song that evokes positive emotions in your viewers.
2. Create a contingency between the competitor's product (CS) and a negative song (US). When listening to the negative song, viewers will experience negative emotions (UR). Since a contingency has been formed between the song and the product, viewers will also experience negative emotions (CR) when viewing the competitor's product in the absence of the song.
3. **CS**: Carbonated beverage
   US: Upbeat song
   CR: Positive emotions
   UR: Positive emotions
   Presenting the carbonated beverage (CS) while playing the popular upbeat song (US) will elicit positive emotions in your viewers (UR). Since the beverage and the song are presented alongside each other, a contingency will form between the two. Eventually, customers will be presented with the beverage in the absence of the song but will still feel positive emotions (CR), making them want to buy your product!
4. Stimulus generalization is the classical conditioning principle at play in this case. In stimulus generalization, stimuli similar to the CS also produce the CR. Stimulus discrimination restricts the range of CSs that can elicit a CR. It uses an extinction procedure to extinguish the CR in which the CS is presented in the absence of the US. In this case, students should also be required to perform another oral presentation in a course that is NOT history. The student would then be exposed to the CS (class presentations) in the absence of the US (the history course that they struggle in). This would eliminate the fearful response that the child naturally has in the history course.
5. Individual answers may vary. Please consult your TA if you require assistance.

Module 3: Classical Conditioning – Review Questions

1. **The US**: A stimulus that automatically triggers a response in the absence of learning.
   The UR: A response, which is automatically triggered by a stimulus in the absence of learning.
   The CS: A previously neutral stimulus that, when paired with the US, a learned contingency is formed between the

two such that presentation of the CS reliably predicts the presence of the US.

The CR: The response, which occurs in response to the CS once a contingency between the CS and US has been learned.

2. Acquisition is the process by which the contingency between a CS and US is formed. The negatively accelerated increasing function suggests that learning a contingency takes many trials. The majority of learning occurs during the early trials. In subsequent trials learning occurs, but never as much as in the earlier ones.

3.
   a) **US**: Flu. **UR**: Nausea. **CS**: New food. **CR**: Nausea.
   b) **US**: Passionate date. **UR**: Joy. **CS**: Date's perfume. **CR**: Joy.
   c) **US**: Panic attack. **UR**: Fearful feelings. **CS**: Airplane. **CR**: Fearful feelings.

4. Stimulus generalization occurs when stimuli similar to the CS also produce the CR. In a classical conditioning experiment a 500Hz tone (CS) is paired with food (US). This leads to salivation (UR) in the participant. Eventually, the presentation of the 500Hz tone (CS) alone will lead to salivation (CR). Once the contingency between the 500Hz tone (CS) and the food (US) has been established, experimenters test for stimulus generalization by presenting the participant with tones of varying frequencies and measuring the salivation (CR). The strongest CR will be elicited by the original 500Hz tone but stimuli similar to the original tone CS (i.e. 600 Hz) will also elicit the salivation CR at similar levels. The more different the tone CS becomes, the less salivation CR is elicited. Stimulus discrimination restricts the range of CSs that elicit a CR. In discrimination training, CR to other CSs is extinguished by repeatedly presenting a CS at the far end of the generalization gradient in the absence of the US. As a result, other tones similar to the original 500Hz still elicit a CR as they did before but the one for which the CR was extinguished does not. For example, extinguishing the CR to the 600 Hz tone will not alter the CR to the 400 Hz tone.

5. Extinction refers to the loss of a CR when the CS no longer predicts the US. Extinction occurs when the CS is repeatedly presented in the absence of the US so that the CS no longer elicits the CR. In the salivation example, the 500Hz tone would be repeatedly presented in the absence of food. Following an extinction procedure the CR to the CS gradually fades. But, following a rest period, the CS is presented again and the CR occurs! This is called spontaneous recovery. Spontaneous recovery suggests that extinction involves a *new* inhibitory learned response. The original learned association between the CS and the US is not actually "unlearned". Instead, extinction promotes a learned inhibitory response that competes with the original contingency.

6. Homeostasis is the process of keeping the internal state of the body constant, including but not limited to temperature, glucose and ion levels. Classical conditioning allows the body to prepare for challenges to homeostasis, such as changes in blood glucose levels after sugar intake.
US: Sugar intake
UR: Insulin release (to counter the rise in blood glucose levels)
CS: Taste of sugary food
CR: Insulin release (to counter what your body thinks will be a rise in blood glucose levels)
Eating something that tastes sweet (CS), suggests sugar intake (US). This usually means that blood glucose levels will soon rise. To counter this rise in blood glucose levels and maintain homeostasis, a CR of insulin release occurs.

7. Because drugs are never taken in isolation, chemical changes associated with drug administration are also signalled by environmental cues. The **drug effects** are the **US** and the **counter-adaptations** which the body undergoes to counter the drug effects and maintain homeostasis are the **UR**. **Environmental cues** specific to the conditions under which the drug is taken are the **CS** (e.g. rooms, smells or people present). A contingency is formed between these environmental cues (CS) and the drug effects (US). Eventually, the presentation of the environmental cues alone will trigger the **CR** of **counter-adaptations** by the body to counter the drug effects anticipated by the body.

Module 4: Instrumental Conditioning – Courseware Exercise

1.
   a) **Name**: Reward Training. **Description**: Presentation of a positive reinforcer. **Example**: Every time Billy picks his jacket off the floor, his mother gives him a cookie.
   b) **Name**: Punishment. **Description**: Presentation of a negative reinforcer. **Example**: Every time Billy leaves his jacket on the floor, his mother scolds him.
   c) **Name**: Omission training. **Description**: Removal of a positive reinforcer. **Example**: Every time Billy leaves his jacket on the floor, his mother gives him a "time out". During the time out Billy must sit alone and he is not allowed to play video games with his brothers and sisters.
   d) **Name**: Escape training. **Description**: Removal of a negative reinforcer. **Example**: Every time Billy picks his jacket off the floor, his mother turns off the classical music playing in the house, which Billy does not like.
   e) Billy's mother may be waiting too long between Billy's behaviour and administering the consequence. To ensure that conditioning is effective, the time between the behaviour (Billy leaving his jacket on the floor) and the punishment (Billy's mother scolding him) should be short; otherwise, Billy may do many other things in between leaving his coat on the floor and the reinforcement and would have difficulty learning which behaviour lead to the consequences.

2.
   a) FR-3
   b) FR-2
   c) Since Carrie will now be picking fewer buckets of strawberries to receive her pay, the run part of the graph, corresponding to the y-axis will become smaller.
   d) FI-1

Module 4: Instrumental Conditioning – Review Questions

1.

	Similarities	Differences
Classical Conditioning	- both involve the formation of contingencies	- contingency is formed between two stimuli - response is involuntary or unconscious
Instrumental Conditioning	- both involve generalization, discrimination and extinction	- contingency is formed between a behaviour and a consequence - behaviour is voluntary or conscious

2. c
3. b
4. b
5. Fixed ratio

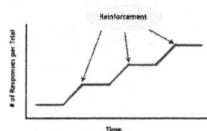

Fixed interval

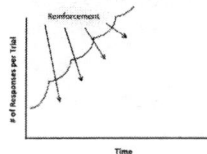

Variable ratio

Variable interval

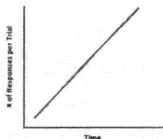

Module 5: Problem Solving – Courseware Exercise

1. Individual answers may vary. Please consult your TA if you require assistance.
2. Individual answers may vary. Please consult your TA if you require assistance. Ps! Check this out! "Chimpanzee Problem Solving": http://www.youtube.com/watch?v=ySMh1mBi3cI

3. Individual answers may vary. Please consult your TA if you require assistance.

Module 5: Problem Solving – Review Questions

1. Deductive reasoning is the process of determining a specific fact from a general theory. Inductive reasoning is the process of determining a general theory from simple facts.
   a. Scientists start with a general theory about the world and use *deductive reasoning* to generate a specific testable hypothesis about the data they expect to obtain. Then through experiments, scientists collect data, using *inductive reasoning* to relate their findings to their general theory in a meaningful way. Therefore scientists generate testable hypotheses using deductive reasoning and interpret collected data with inductive reasoning.
2. The confirmation bias is our tendency to seek out information that supports our hypothesis. For example, when buying an item of clothing on sale, people often look for similar items in more expensive stores, to convince themselves that they received a good deal.
3. The availability heuristic is our tendency to make decisions based on the information that is most quickly available to us. For example, after watching Shark Week on the discovery channel, you believe that the likelihood of encountering one in the ocean is much higher than it actually is. On vacation, you are afraid to swim.
4. The representativeness heuristic is our tendency to assume that what we are seeing is representative of a larger category we have in our mind. For example, you run into an acquaintance in line for lunch at the cafeteria. Although you can't seem to remember where you know them from, you notice that they are carry goggles in the side pocket of their backpack. You then assume that you've met them in your chemistry lab, and proceed to talk about your most recent assignment (of which they have no idea!)
5. Reliability measures the extent to which repeated testing produces consistent results. Validity measures the extent to which a test measures what the researcher actually claims to be measuring.
6. James uses the availability heuristic to make this decision not to study. When James thinks of students who get A+ grades, he immediately thinks of his friends who do not study as this information is most what is most readily available to him.
7. Answer: d
   No one knows! The above options are possibilities that have been proposed but the primary explanation for the Flynn Effect remains an open question to be answered.
8. Both genetics and the environment play a role in determining IQ. See Unit 4, section "Genetic and Environmental Conditions" for examples.
9. c
   Neither child will be able to answer the question correctly. It

requires the kind of abstract reasoning that develops later, in the formal operational stage (age 11+).

Module 6: Language – Courseware Exercise

Animal example	Regular?	Arbitrary?	Productive?	Is it language?
Waggle Dance	Yes	No	No	No
Bird Song	Yes	No (many bird calls are highly suited for their purpose)	No	No
Washoe	No	Yes	Yes	No
Sarah	Yes	Yes	No	No
Kanzi	No	No	Yes	No
Human Speech	Yes	Yes (with the exception of onomatopoeic terms)	Yes	Yes

Module 6: Language – Review Questions

1. c
2. b
3. Milestones of Language Development

Age	Milestone
12 weeks	Makes cooing sounds
16 weeks	Turns head towards voices
6 months	Imitates sounds
1 year	Babbles
2 years	Uses 50-250 words; uses two word phrases
2.5 years	Vocabulary > 850 words

   a. The language explosion occurs between about one and a half to six years of age. At this time, vocabulary increases rapidly and most children have mastered the major aspects of language.
4. In spoken language, segmentation is the process of determining when one word ends and the next begins.
   a. An infant's proficiency at speech segmentation predicts their language ability in the future. There is a strong positive correlation between early speech segmentation and expressive vocabulary at two years of age. Thus, children with large expressive

vocabularies have often demonstrated good speech segmentation skills as infants.
5. Universal phoneme sensitivity is the ability of infants to discriminate between any two sounds, including sounds from non-native languages.
   a. Evidence for the existence of universal phoneme sensitivity comes from an experiment in which researchers compared English speaking adults, Hindi speaking adults, and infants from English speaking families, on their ability to discriminate between two different /t/ sounds. These sounds are present in the Hindi language but not in English. Adult English speakers performed worst on this task, whereas infants from English speaking families performed almost as well as the Hindi-speaking adults. Thus, infants can discriminate non-native sounds that are absent from the language of the culture in which they are being raised.
6. Social Learning Theory:
   Supporting Evidence: Without exposure to adequate sources of language, children fail to develop language skills. (Eg. Genie)
   Opposing Evidence: Children's language development is far too rapid and complex to be driven by imitation and repetition alone. In addition, children combine words in novel ways that have never been modeled or reinforced. Finally, children make language errors that are not heard in adult speech (eg. overextensions and underextensions).
   Innate Mechanism Theory:
   Supporting Evidence: Children who have been taught to lip read can spontaneously sign, and use innate and automatic grammar when doing so. In addition, young infants show neurophysiological responses to the first language they are exposed to, indicating that their brains are pre-wired to adapt to the sounds and their meanings that are present in their environment. Finally, infants prefer to listen to speech sounds (as opposed to non-speech sounds) indicating that they have an innate predisposition to expose themselves to language.
   Opposing Evidence: None listed in lecture.

Module 7: Categories and Concepts – Courseware Exercise

1. Prototype theory could explain the discrepancy in ratings for the object "palm tree". Prototype theory suggests that we categorize objects by comparing them to an internal representation of the category, called a prototype. Prototypes are the average, or "best", member of a category and are formed through personal experience. As a result, prototypes are subjective; they depend on the objects each person has previously encountered and averaged together. Since palm trees do not grow in Canada, they have would not weigh as heavily in to the averaged Canadian prototype of a tree.

2. The maple tree. It has both high typicality and membership ratings, therefore, not only is it considered a member of the category "trees", it is considered to be a typical, or average, member of the category "trees".

3. Exemplar theory can also explain these results. It suggests that instead of storing *one* internal average of a category, we store *every experience* of that category over our entire lifetime as exemplars. When categorizing an object we search through our library of exemplars to find one similar to the current object. Once found, we can then identify the current object as a member of the same category. According to exemplar theory, It's not that the maple tree looks more like their average tree prototype, but that the Canadian subjects have more maple exemplars than palm exemplars. Thus, they are able to retrieve a maple tree exemplar much more quickly and rate it higher in membership and typicality.

Module 7: Categories and Concepts – Review Questions

1. **Classification**: Categorization allows you to treat objects that appear differently as belonging together. *For example, no matter if they're green or purple, they're still grapes!*
   Understanding: Categorizing a scene allows us to understand and react appropriately. *For example, while sitting on a crowded subway, you notice that the elderly woman in front of you is having trouble balancing while standing in the moving car. You immediately understand that she is struggling and offer her your seat.*
   Predicting: Categorizing your current experience and comparing it to similar experiences in memory, allows you to make predictions about your current situation. *For example, when cat-sitting for your neighbour you know not to pull its tail as it will get angry and scratch you.*
   Communication: Many of the words in our language refer to categories or concepts! Using the category names allows for efficient communication. *For example, when conversing with one another in the emergency room, doctors and nurses understand one another when using jargon you have no understanding of because their jargon is not part of their everyday vocabulary but is not a part of yours.*

2. It is not possible to describe categories with a set of rules. When asked to define a category with a list of necessary and sufficient rules, it is difficult to properly include and exclude items in the category. In addition, it is especially difficult to devise simple rules to define more abstract categories.
   a. Experiments on the use of rules tell us that for simple categories we are quite susceptible to the illusion of the expert. Since categorization is easy for us, it must have a simple rule! In actuality, categorization cannot be established through a set of rules. Instead, scientists believe that humans have an internal representation of categories that is independent of the rules we try to define.

3.

	Prototype Theory	Exemplar Theory
Similarities	Used in categorization. Compares a new experience to an internal image.	Used in categorization. Compares a new experience to an internal image.
Differences	Suggests that we store one internal average of a category.	Suggests that we store our entire lifetime's worth of experiences.
How does the theory explain the categorization of an "orange" into the category "fruit"?	Prototype theory suggests that we categorize objects by comparing them to an internal representation of the category called a prototype. Prototypes are the average or "best" members of a category and are formed through personal experience. Thus all of our encounters with oranges have been averaged into our prototype of "fruit".	Exemplar theory suggests that we store our entire lifetime of experiences as exemplars. When viewing an orange, we search through our library of exemplars to find one similar to it. Since we come across oranges frequently, we are able to categorize it quickly as a type of fruit.

4. Children as young as three are able to understand general categories as well as category membership. In addition, children also understand the innate properties of a given category.
   a. Evidence that children can apply categories comes from their ability to determine category membership. In the module, Katie learns that her dog likes treats and applies this knowledge to other breeds of dogs as well. She knows that since they are members of the same category, they share similar characteristics. Evidence that children understand categories comes from their understanding of the innate properties of categories. In lecture, its is noted that children understand that you can change the nature of a machine, but that you cannot change the nature of an animal.

5. Evidence that animals can apply categories and concepts comes from research on baboons. Using classical conditioning procedures, baboons have learned to classify objects as either food or non-food, and can do so with 90% accuracy. In addition, baboons are also capable of more abstract categorization and can place objects into the categories "same" and "different" with a high level of accuracy as well.

Module 8: Attention – Courseware Exercise

1. The spotlight model of attention cannot explain this situation. The spotlight model of attention suggests that humans have an attentional spotlight that focuses on part of the environment at one time to the exclusion of the rest of the visual scene. Attention is consciously directed across the visual scene, with objects falling within the spotlight being processed faster and preferentially. Debbie was not consciously attending to the part of the visual scene in which the red car was in, and thus, it was not in her attentional spotlight and should not have be processed.

2. This situation cannot be explained by Broadbent's single filter model of attention. The model suggests that an attentional filter selects important information to be processed on the basis of physical characteristics and allows only that information to continue on for further processing. Information that does not pass through the physical filter is completely eliminated and unavailable for further analysis. In this case, red cars are not relevant to Debbie's physical filter and thus all should be rejected for further processing.

3. This situation can be explained using Triesman's dual filter model of attention. The model suggests two filters. The first filter attenuates unimportant objects based on physical characteristics, but still allows everything through. The second filter determines what is important based on semantic meaning. Although red cars have been attenuated and are less likely to make it through the second filter, the red Thunderbird has semantic meaning to Debbie and thus makes it through.

Module 8: Attention – Review Questions

1. Automatic processes of attention are involuntarily triggered by external events and operate in a fast, efficient and obligatory manner. Controlled processes however, are slow and effortful. They voluntarily and consciously guide attention to objects of interest.
   a. Because they require less cognitive effort, automatic processes occur quite quickly. They also allow you to process salient information. In addition, frequently practiced tasks can become "automatic" as a result of learning and thus may require less effort over time.
   b. Using controlled processes is advantageous because it allows one to participate in "goal-directed behaviour".

2. The spotlight model represents an area of enhancement of the visual scene. The "spotlight" area receives enhanced attention, allowing you to react more quickly to cues inside it and identify things within it more accurately. You are still paying attention to things outside the spotlight, but your reaction to them is slower and less accurate unless you move your spotlight to a new part of the visual scene.

3. The spatial cueing paradigm is an experimental procedure that allows researchers to manipulate the attentional

spotlight. In the paradigm, subjects stare at a screen on which there are three boxes. Participants are asked to focus their attention on the middle box (beside which are boxes on the left and right sides). A target will then appear in either the left or right box and subjects are told to identify it as quickly as possible. However, before the target appears in one of the boxes, a potential target box briefly flashes, serving as a cue for the subject's attention. The real target will then follow, either in the cued or uncued location.
   a. In Experiment #1, the cue accurately predicts where the target will appear for the majority of trials and so it demonstrates controlled processing. Consciously controlled shifts of attention can lead to faster responses to targets that appear in the location indicated by the cue than to those that appear opposite to the location of the cue.
   b. In Experiment #2, the cue accurately predicts where the target will not appear for the majority of trials. This demonstrates controlled processing again because the participant will chose to ignore the flash and bring their attention to where they expect the target to be. This will lead to faster responses to targets that appear in the location opposite to the cue than to those that appear in the same location as the cue.
   c. In Experiment #3, the cue provides no predictive information about where the target will appear and so it demonstrates automatic processing. Researchers find that response time occurs more quickly when the target is correctly cued. When the target occurs in the cued location, the subject's attention is already there and thus reaction time is quicker. When the target occurs in the uncued location, the subject must shift their attentional spotlight from the location of the cue to the location of the target, and thus reaction time is longer. These results suggest that the difference in target detection is governed by automatic control of attention

4. Broadbent's single filter model suggests that an attentional filter selects important information to be processed on the basis of physical characteristics and allows only that information to continue on for further processing. Information that does not pass through the physical filter is completely eliminated and unavailable for deeper analysis. In contrast, Triesman's dual filter model of suggests two attentional filters. The first filter attenuates (or lowers the weight of) unimportant objects based on physical characteristics, but still allows everything through. The second filter determines what is important based on semantic meaning and chooses what information will be attended to. Information that does not get through this second "semantic filter" is then discarded.
   a. "Breakthrough" is a phenomenon that occurs when people remember unattended information. It is

relevant, as its existence led to the revision of Broadbent's single filter model and the creation of Treisman's dual filter model, in which the additional semantic filter explains why breakthrough can occur.

5. In the Stroop Task, participants are presented with a colour word and asked to name the ink-colour in which the word is presented. Items can be congruent (matching word and colour dimensions) or incongruent (mismatching word and colour dimensions). Researchers measure how long it takes for a participant to correctly identify the ink colour while ignoring the word dimension. Performance is found to be faster on congruent items.

   a. The Stroop task requires you to attend to information in the task relevant-dimension and ignore information on the task-irrelevant dimension. The difference in performance between congruent and incongruent trials acts as an empirical measure of processes involved in selective attention. On congruent trials, the word and its colour match so that reading facilitates colour naming performance. On incongruent trials, the word and its colour do not match and reading interferes with colour naming performance.

6. In a visual search task, subjects look for an object amongst an array of distracters.

   a. In a feature search, the subject looks for one particular feature to identify the target. For example, finding a square in a group of circles is a feature search. In a conjunction search, the subject looks for two or more features to identify the target. For example, finding a green T in a set of red and green Is and red Ts is a conjunction search.

   b. A visual search that proceeds rapidly is called a pop-out effect. A pop out effect occurs quickly, regardless of set size.

   c. Contextual Cueing is when the expected position of the target serves as a cue to decrease the difficulty of a visual search. It is relevant; making search tasks both quicker and easier.

Module 9: Memory – Courseware Exercise

Individual answers may vary for this activity. Please consult your TA if you require assistance.

Module 9: Memory – Review Questions

1. Ebbinghaus operationally defined memory as a serial learning task. Using himself as a subject, he memorized word lists and suggested that each word in the list acted as a cue to trigger the memory of the word that followed. Ebbinghaus noticed that his ability to recall words was highest immediately following learning and that over time he was able to remember fewer and fewer words. This lead

him to create the forgetting curve which describes the increasing rate of memory failure over time.

   a.

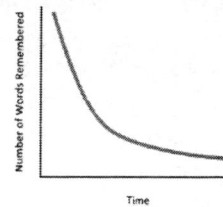

2. Chunking refers to the reorganization of information into meaningful packets. Chunking allows more information to be stored in short-term memory. Because it has been observed that only 7+/-2 items can be held in short-term memory at a time, chunking helps store more information.

3. The Multi-store model assumes that memory is composed of both short and long-term storage systems. According to the model, incoming perceptual information is first stored in a short-term memory buffer. This information in short term memory is only available for current tasks, and is not stored permanently in the memory system. Important information encoded in short-term memory can be transferred to the long-term memory system for more permanent and long-term storage.

4.

   a. Recency will be affected by talking on the phone before your test. Although you studied right up until leaving for the midterm, talking to your friend will have used up your short-term memory capacity, thus diminishing the effect of recency on recall for test information.

   b. Primacy is responsible for the fact that your TA remembers most clearly the first names. Your TA was able to rehearse the names at the beginning of the game the most, enhancing recall of these names as they are rehearsed enough to transfer into long-term memory stores.

   c. Primacy will be affected by this manipulation. Those who are able to view the words for 2 minutes will have better recall for those words at the beginning of the list because they had more time to rehearse and transfer these items into long-term memory. Those who only had 30 seconds to view the list of words would not have had the same opportunity to rehearse and have these words stored in long-term memory.

   d. Recency will be enhanced. Your friend will have better recall for the last items he studied as these will be able to remain in his short-term memory because your friend was not distracted by conversations before the exam which would use up his limited short-term memory capacity.

5. Encoding specificity is the finding that recall is better when retrieval takes place in similar conditions to encoding. Encoding specificity suggests that both our internal and external environments affect our memory. Specific aspects

of our original experience influence our memory performance in the future by acting as cues for the event or item being recalled. Memory performance thus depends on both how we encode items *and* the encoding context. In a study by Godden & Baddelley, SCUBA divers were asked to encode a list of words while on land or underwater. A follow up test was then preformed in either the same or a different encoding context. Results demonstrated that subjects were better able to remember items from the list when they were in the same original encoding context during their follow up memory test.

6. Please refer to module 8 slide 57-60 for a description of the experiment. This experiment suggests that misattribution can occur. The false fame effect occurs as a result of processing fluency of a non-famous name improperly attributed to fame.

Module 10: Forming Impressions – Courseware Exercise

1. Individual answers may vary. If you require assistance, please consult your TA.

2. Sarah is making use of the availability heuristic. Sarah assumes that Monica will be more fun to go to the concert with, but it is possible that this assumption is incorrect. Sarah is using the relative availability of positive memories from her psychology class compared to her physics class in order to rate Monica as a more fun friend than Maria. However, Maria could be a very fun person outside of physics class!

Module 10: Forming Impressions – Review Questions

1. Degree of choice, expectation, and the intended consequences of behaviour.

2. **Consistency**: Does this individual usually behave this way in this situation? We can assume the answer is yes from the question. It is assumed that he behaves the same way each time he gives a speech as a student's union presidential candidate.
   Distinctiveness: Does the individual behave differently in different situations? Yes. Because your friend seems to know him and thinks he is a nice, humble guy, it is likely that he usually behaves differently than this seemingly conceited fashion in other situations.
   Consensus: Do others behave similarly in this situation? Yes. Others who are running for president would also tell of their accomplishments and community involvement.
   Based on the answers to these questions, you can assume that the male student's behaviour is situational, not dispositional.

3.
   a. The Fundamental Attribution Error (FAE) is our tendency to overvalue dispositional factors for the observed behaviours of others while undervaluing situational factors.

   b. d
   c. Generally, FAE is diminished in collectivist societies (such as China or India) where there is less focus on individual behaviour and more on relationships and roles in society. Western societies; however, are generally individualistic and focus on the individual and are thus more likely to attribute behaviour of others to dispositional causes over situational causes.

4. Using the representativeness heuristic one can classify people depending on how well their behaviour fits with a certain prototype. Using the availability heuristic, one can decide how likely it is for a person to possess a certain trait by considering how easy it is to come up with examples of that person acting a certain way.

5. **Proximity**: Refers to both physical and functional difference. You are more likely to become attracted to a person if you see and interact with them often.
   Familiarity: We tend to feel more positively about things that we have seen before.
   Physical Attractiveness: We presume that "what is beautiful is also good"
   Others Opinions: You like people...who like you!

Module 11: Influence of Others I – Courseware Exercise

1. Groupthink was at the root of NASAs problematic decision-making process. Groupthink occurs when group cohesion is so strong that it overrides realistic appraisals of reality and alternative options.

2. Your solution should include the following: be impartial, use critical evaluation, allow group members to disagree, subdivide the group into smaller ones, call a second chance meeting before implementing a final decision.

Module 11: Influence of Others I – Review Questions

1. Triplett observed that cyclists raced faster when competing with each other in a group than when they raced against the clock in individual time trials. This phenomenon, in which the presence of co-actors or an audience leads to enhanced performance, is called social facilitation.
   a. The presence of others does not always improve performance. Sometimes it actually hinders it!
   b. Zanjonc suggested that the presence of others increases your arousal. The way this heightened arousal affects your performance depends on the specific task to be performed. For simple tasks in which you are well-practiced, performance is enhanced. For complex tasks in which you have little practice, performance is hindered.

2. Bandura's experiment is important as it demonstrated social learning theory; you learn appropriate behaviours by modeling or imitating the behaviours of others.

3. Individuals sat alone in a dark room, observing a stationary light on a wall. Despite it being immobile, the autokinetic effect made the light look like it had moved a small amount.

Participants were asked to estimate how much the light moved. The next day, participants sat in the same conditions, except this time two other subjects were present in the room. All participants were asked to provide their estimates of how much the light moved out loud in front of the group. Following a few more days in the same conditions, participants responses gradually converged, despite the fact that they all had given different estimates on the first day. Even when a confederate of the experimenter sat in with the subjects and gave a very large estimate of the light's movement on the wall, the group's responses still converged to a large number, incorporating this confederate's response as well.

4. Asch's participants were seated in a room with a group of confederates. They were told that they would see one sample line and three comparison lines and that then they would be required to identify which of the comparison lines matched the standard. As the experiment proceeds, confederates start to agree on answers that are clearly wrong. When it is the actual subjects turn to answer, the majority also conformed to the incorrect response! This experiment demonstrates the normative function.

5. Deutsch and Gerard's modification to Asch's experiment demonstrates the comparative function; we look to the actions of others to provide information about an ambiguous situation. Subjects performed the same task as in Asch's experiment, but this time in cubicles with anonymous identities. Before a subject was to respond, lights lit up on their screen that indicated the results of other participants in the experiment. Although anonymous and free of the potential to be ridiculed, subjects sill went along with the rest of the group, even on incorrect answers. Conformity occurred mostly when answers were less clear. Because subjects doubted their own perceptions, they went along with those of the group suspecting that others may be right.

6. Both the normative and comparative functions outline the roles of others in setting standards for our conduct. In the normative function, conduct is based on fear of rejection whereas in the comparative function, conduct is based on the actions of others in an ambiguous situation.

7. Group polarization occurs when group decision-making strengthens the original inclinations of the individual group members. If the individuals were originally inclined to a low level of risk, the group would make an even less risky decision; however, if they individuals were originally inclined to a high level of risk, the group decision would be even riskier than the individuals would have tolerated. The risky shift refers to the idea that some group decisions have riskier outcomes than when those same decisions are made by individuals alone.

8. Darley and Latane's experiments demonstrated collective ignorance. When each individual in a group sees nobody responding in a given situation, they conclude that the situation is not an emergency. When subjects were alone,

they were more likely to report smoke seeping in a room than if others were in the room and this likelihood of reporting the smoke decreased as the number of people in the room increased

a. Darley and Latane's experiments also demonstrated diffusion of responsibility.

In deciding whether we have to act, we determine that someone else in the group is more qualified. To break diffusion of responsibility, it helps to be direct. Choose an individual out of the group and place the responsibility on them.

9. Social loafing is a special case of diffusion of responsibility, in which individuals are less motivated when working in a group than when working alone.

Module 11: Influence of Others II – Courseware Exercise

1. The communicator should have high credibility but also be close in age to a university student. Perhaps a young, attractive PhD student studying psychology.

2. Your audience is intelligent (they are psych students aren't they!?), therefore use a central appeal. Because the students may not initially agree that they should buy the textbook, you should present a well reasoned and factual two-sided argument.

3. Your audience is university students. The subpopulation of students that you are appealing to are those studying psychology.

4. In this campaign, you should attempt to have the students overjustify the behaviour of buying the textbook. Because their behaviour (buying a more expensive textbook) doesn't match their attitude (feeling anxiety over spending more money on a tight budget), you want them to feel that the textbook is so great that paying the extra 10% for it was worth it and that they are glad they didn't buy a far inferior textbook. The overjustification is: "well this textbook is much better anyways" to relieve cognitive dissonance.

Module 11: Influence of Others II – Review Questions

1. Please refer to lecture 17, slide 4 for a full description of the experiment.
   a. It was found that the majority of subjects continued on to the end of the experiment, delivering a shock labelled "high shock voltage" to a patient with a heart condition.
   b. The prestige of the experimenter and environment as well as the physical separation between the teacher and the learner could have contributed to the results. In addition the proximity of the experimenter to the learner could have also played a role.
   c. Many lessons can be learned from Milgram's experiment. We learn that humans have a very strong tendency to obey authority. In addition we learn that we do not always accurately judge how we

would behave in a given situation. Finally, this experiment also makes us deeply consider ethics when conducting experiments on human subjects.

2. Subjects who were given $1 to lie about the experiment had cognitive dissonance because their attitude (the experiment was boring) did not line up with their behaviour (telling the next subject the experiment was fun). To relieve this dissonance, they need to make their attitude and behaviour line up. Because their behaviour could not be changed (they had already lied), they had to change their attitude and therefore rated the experiment as being more fun. Those subjects given $20 on the other hand, could justify their lying behaviour by saying that they only lied for the money, thus relieving cognitive dissonance without changing their attitude toward the experiment. As a result, their attitude that the experiment was boring remained and they rated it lower.

3. Please see lecture 17, slide 29 for details.

   a. This study did not run the entirety of its two-week course. Instead, it was halted after only 6 days as subjects assigned as guards were acting sadistically towards subjects assigned as prisoners.

      b. The experiment demonstrates the power that circumstance and assigned role can have on human behaviour. It also demonstrates how de-individuation may result in dangerous behaviour.

4. The most persuasive communicators are those with high credibility. In addition, both physical attractiveness and how relatable or similar the communicator is also play a role. For decisions of lifestyle choices, a more similar communicator is more persuasive while for factual decisions, a more credible communicator is most persuasive. Persuasive communicators also speak in a more straightforward, concise manner and maintain better eye-contact.

5. It depends on the audience! A two-sided message is better if your audience initially disagrees with your opinion. A one-sided message is more effective if your audience is already in favour of your message.

6. In a central appeal, a well-reasoned, two-sided argument is presented. This type of appeal often works best when addressing an academic audience. In a peripheral appeal, a well presented, easy to understand message is presented. Peripheral appeal is often more effective for unintelligent audiences.

7. The foot in the door technique refers to a situation in which a gradual escalation of demands increases obedience. Just as the foot in the door technique results in a gradual escalation of demands, the low-ball technique does too! However, In the case of the low ball technique, compliance is secured at a smaller cost, only later to reveal additional costs while making the original decision seem irreversible.

Foot in the door: For example, suppose a vacuum salesman comes to your door. He convinces you that his vacuums are the best and you agree to buy the basic model for $300 even though you already have a good vacuum and weren't planning on getting a new one. Once you sit down to sign the papers, he shows you the features that you are missing out on by only buying the basic model and you now agree to buy the upgraded model for $450. Finally, he demonstrates the use of some additional vacuum attachments to make cleaning much easier and you agree to pay another $150, bringing your total to $600.

Low Ball: The same vacuum salesman comes to your door and easily persuades you to purchase a vacuum for only $300! When you sit down to sign the papers, he informs you that you will need to pay additional charges of $100 for a customer-care package, $50 for necessary vacuum bags, $50 for shipping and $100 in duty charges as the vacuum will be shipped from the USA. Because he secured your compliance at the a lower rate at the beginning, you now feel that your original decision is irreversible and agree to pay $600 for your new vacuum

GLOSSARY OF TERMS

A

Above Average Effect: The tendency for an individual to assume that they are better than the average person, and that the challenges and failures of the average person are not ones they will experience. *E.g. Despite the increased awareness about the risks of drunk driving, some individuals may still drive when intoxicated, claiming that, "Other people get into accidents. I won't." (Forming Impressions)*

Accent: A manner of pronunciation unique to an individual, geographical location, or nation. It occurs both between and within languages. It may be influenced by location and dominant speech patterns in the environment in which an individual is raised. *(Language)*

Acquisition: In classical conditioning, it is when a contingency is formed between a CS (conditioned stimulus) and US (unconditioned stimulus) following repeated presentations of one after the other. *E.g. A researcher repeatedly rings a bell before presenting food for a dog. Now, when the dog hears the bell, he salivates in anticipation of food. (Classical Conditioning I)* In instrumental conditioning, it is when a contingency has been learned between a behaviour and a consequence. *E.g. When you were a child, you learned that being polite lead to positive consequences and now, as an adult, you are still polite in hopes of positive consequences. (Instrumental Conditioning)*

Actor/Observer Effect: A proposed explanation of the fundamental attribution error, stating that an individual is more likely to make the fundamental attribution error when determining causes of others' behaviour rather than their own. As the actor, they are better aware of the situational factors contributing to their behaviour; but when observing others, they only have the current situation and assume that what you see is typical of that person. *E.g. Drivers may be more likely to attribute their own speeding to situational factors such as being late while attributing the speeding of others to dispositional factors such as they are bad drivers or aggressive people. (Forming Impressions)*

Addiction: In terms of classical conditioning, when a stimulus, often a drug (US), elicits some sort of drug effect (UR). Over time after continued use, the environment in which the drug is taken in (CS) will be able to trigger compensatory responses (CR) which prepare the body for the effects of the drug, attempting to maintain homeostasis in the body. *(Classical Conditioning)*

Agreeableness: One of the personality traits in the "Big Five" model. This trait reflects an individual who is warm, compassionate, polite and caring rather than confrontational or hostile. *(Personality II)*

Anal Stage: The second of Freud's psychosexual stages of personality development that occurs from ages 1 - 3. During this stage, the child learns that controlling bladder and bowel movements can be pleasurable activities. *(Personality I)*

Anima: According to Jung, the anima is both an archetype and complex that men have. As an archetype, the anima is every man's unconscious, instinctive idea of femaleness. As a complex, it is the thoughts and feelings that a man rejects from consciousness because they are seen as feminine. *(Personality II)*

Animus: According to Jung, the animus is both an archetype and complex that women have. As an archetype, the animus is every woman's unconscious, instinctive idea of masculinity. As a complex, it is the thoughts and feelings that a woman rejects from consciousness because they are seen as masculine. *(Personality II)*

Arch of Knowledge: A diagram used to illustrate the different mechanisms of reasoning in problem solving. Specific facts are found at the base of the arch, while general theories are found at the top of the arch. The general theories govern how all the facts at the bottom are related using deductive reasoning. More specific facts can be determined through experimentation, which may contradict or supplement the general theories. Therefore, broader theories can be altered through inductive reasoning. *(Problem Solving)*

Archetype: According to Jung's theory, basic human instincts shared by all human individuals in the collective unconscious, that allow us to organize and interpret experiences in certain ways. Archetypes are unconscious so they are seen only through projections throughout history in literature, myths, religions, etc. *E.g. Some archetypes included themes such as hero, social conformity, birth and rebirth. (Personality II)*

Asch's Stimuli: An experiment performed by Asch, in which a participant (among confederates) was required to identify which of three lines matched a standard in length. If all the confederates in the experiment gave an incorrect response, the participant would conform socially and agree with the incorrect answer despite the obviousness of the mistake. *(Influence of Others I)*

Attraction: In attraction research, this term refers to having a good impression of someone or something and desiring their company. It does not refer exclusively to sexual attraction. *E.g. To say you are attracted to your best friend whom you see often is to say that you have a good impression of that person. (Forming Impressions)*

Attribution: Judgments that tie together causes and effects. *E.g. If you see someone you recognize, the effect is the feeling of fluency or a feeling that the person is familiar. You begin to investigate the cause for this effect, or where you know the individual from. (Memory)*

Audience: A group of people watching an individual perform a task, which can the performance of the actor through social facilitation. *E.g. During a concert recital, a famous cello player may find that he performs better in front of a large audience than when rehearsing in front of his cat. (Influence of Others I)*

Automatic processes: When external events involuntary "capture" one's attention. *E.g. When hearing a loud bang, one may automatically direct their attention to the source of the noise, even if they did not intend to attend to the noise. (Attention)*

Autoshaping: When a behaviour can be learned without explicit training by an experimenter. *E.g. When the cat is in the puzzle box trying to escape, its random behaviour will eventually lead it to pull the rope without any explicit instructions to do so. Over time, it will be able to consistently do this on its own. (Instrumental Conditioning)*

Availability Heuristic: The tendency to make decisions based on the information that is most quickly available to us. *E.g. You may believe that typical lottery winnings are in the millions of dollars because these are most often publicized in the news; however, most winnings are actually a few dollars. You believe this because this information is most readily available to you. (Problem Solving)*

B

Behaviourist Perspective: The perspective in psychology that asserts that overt behaviour is the *only* valid means of measure in psychology. According to this perspective, the mind is an off limits "black box" and what takes place within should be considered outside the domain of science. *E.g. researchers adopting the behaviour perspective may design carefully controlled experiments where they manipulate the environment and observe the effect on overt behaviour. (Levels of Analysis)*

Behavioural Approach: An approach to personality, which states that an individual's personality simply consists of their behaviour. Theorists that take this approach are not interested in the psychic structures of the mind or underlying personality traits or thoughts. *(Personality II)*

Between-subjects design: An experimental design in which an independent variable is manipulated across *separate* groups of participants, in order to observe the change in the dependent variable. One group of subjects receives the experimental manipulation while the other acts as a control group. Each participant or group of participants will take part in only one of two or more conditions. *E.g. When testing the effects of caffeine on alertness, researchers test ability to recall lecture information in students who drink coffee versus those who do not drink coffee. (Research Methods I)*

Big Five: The most commonly accepted personality factor model which reduces personality factors to five personality traits which encompass most other personality traits: openness, conscientiousness, extraversion, agreeableness, and neuroticism (OCEAN). *(Personality II)*

Biological Perspective: The perspective in psychology that focuses on the physiological mechanisms that underlie thoughts and behaviour, including the structure and function of the brain, how hormones and neurotransmitters and genetic factors contribute to behaviour. *E.g. A psychologist adopting this perspective may examine how the levels of testosterone in the brain affect aggression. (Levels of Analysis)*

Blinding: When participants do not know whether they belong to the experimental or control group, or which treatment they are receiving. *E.g. When comparing the effect of drinking caffeinated or decaffeinated coffee before bedtime, the participants would not know which type of coffee they are drinking. (Research Methods I)*

Bobo Doll Experiment: An experiment performed by Bandura, in which a child would model an adult's behaviour when playing with a toy called a bobo doll. If the adult exhibited aggressive play, the child modeled the same behaviour. If the adult exhibited passive play, the child modeled the same behaviour. No explicit reinforcement was required. *(Influence of Others I)*

Broadbent's Single Filter Model: A model of attention in which a filter selects important information on the basis of physical characteristics and allows only that information to continue on for further processing. *E.g. When looking into a crowded room for a friend wearing a blue shirt, all people wearing shirts of a different colour would not pass through your filter. Your filter would allow you to only focus on the people wearing blue to help direct your attention toward locating your friend. (Attention)*

Bystander Effect: A phenomenon in which individuals do not offer help to a victim in an emergency situation when they think others are present. *E.g. If an individual views someone being harassed when many other people are present, they are less likely to offer help than if they were the only person around. (Influence of Others I)*

C

Categorization: The process by which objects or ideas are recognized, understood, and differentiated by grouping them into different subtypes, each of which has some underlying classification criteria, with some sort of general rule. *(Categories and Concepts)*

Causation: The suggestion that one variable caused an effect to occur in another variable. If two variables have a strong relationship or are strongly correlated, this does not mean that the relationship between the variables is causing the observed effect. *E.g. A researcher may observe a positive correlation between hat sales and ice cream sales. However, hat sales do not cause more ice cream sales. They both may be caused by a confounding variable, an increase in temperature due to warmer summer months. Therefore, an increase in temperature is causing hat sales and ice cream sales to increase. (Research Methods II)*

Central Appeal: Well-reasoned, factual, two-side arguments which are effective for intelligent audiences. *E.g. When trying to persuade a group of academics, factual and two-sided arguments will be a more successful form of persuasion. (Influence of Others II)*

Chunking: Reorganizing information by grouping smaller items into more larger packets in order to allow more information to be remembered. Chunking can allow individuals to remember more significant information within the short term memory capacity of 7 +/- 2 chunks. *E.g. If each item is actually a meaningful combination of 3 numbers, then approximately 21 numbers can be remembered within those 7 items. (Memory)*

Classical Conditioning: Learning a contingency or association between a particular signal and a later event that are paired together in time and/or space. *E.g. The first famous example involves Ivan Pavlov's dog, who learned a contingency between the ringing of a bell and the arrival of food. When the dog heard the bell, it would begin to salivate. (Classical Conditioning)*

Co-Actor: Another individual performing the same task as someone, often influencing performance through social facilitation. *E.g. If an individual is bowling, his or her performance may be increased if they are competing against other bowlers. (Influence of Others I)*

Cognitive Perspective: The perspective in psychology that argues that internal mental processes are necessary to fully understand behaviour. This perspective is not necessarily concerned with describing the mind in terms of the physiology of the brain, but uses models to construct abstract representations of how the mind functions. *E.g. the Single Filter Model for attention is an abstract model used to represent how cognitive psychologists believe attention works as a mental process. (Levels of Analysis)*

Cognitive Approach: An approach to personality that states that central to personality are internal thoughts, interpretations and understanding, which determine how an individual feels and behaviours . *(Personality II)*

Cognitive Dissonance: A psychologically uncomfortable state that occurs when a person's attitude and behaviour are inconsistent. Cognitive dissonance must be resolved by changing one's attitude as one's behaviour cannot change after the fact. *E.g. Christa exercises regularly and is a huge advocate of healthy living. One day, when at home by herself, she orders two large pizzas and finishes them. In order to resolve the resulting cognitive dissonance, she convinces herself that sticking to a strict, healthy diet is not one of her main priorities. (Influence of Others II)*

Collective Unconscious: According to Jung's theory, the collective unconscious is an ancient part of the human mind that forms the biological basis of human nature. It contains basic human instincts, called archetypes, which are shared by all humans. *(Personality II)*

Communicator: The person persuading another. The higher the credibility or attractiveness of the communicator, the more likely they are to persuade another individual. Additionally, a communicator who has more in common with an individual is more likely to persuade them. *E.g. An attractive communicator of similar age to yourself may be the most persuasive communicator in persuading you to buy a new car. (Influence of Others II)*

Comparative Function: When individuals use information from group members to provide information in ambiguous situations and to make decisions. *E.g. If there is a sign that explicitly says "The water may be unsafe – swim at own risk", and yet there are others in the water, an individual may use the information from the group to make their own decision to go swimming. (Influence of Others I)*

Complexes: According to Jung's theory, the collection of memories and emotions are connected by a common theme, which is found in the personal unconscious and helps form one's personality. Complexes typically correspond to an archetype. *E.g. Some people have an inferiority complex and so they spend a lot of psychological energy on ideas, feelings and behaviours related to feelings of inferiority. (Personality II)*

Compensatory Responses: A natural bodily response to maintain homeostasis by compensating for an oncoming change. *E.g. If an individual enters a very hot room, their body temperature will increase. They will then begin to sweat to cool their skin through the evaporation of water, thus helping lower their body temperature back to homeostatic levels. (Classical Conditioning)*

Concrete Operational Stage: The third of Piaget's stages of development, from ages 7 – 12. It is the stage at which the child has mastered all the tasks from the preoperational stage. However, they have difficulty thinking in abstract terms, which means that they have difficulty reasoning using hypotheses. *E.g. A child in this stage might understand the conservation of mass but couldn't understand something more abstract, such as algebra (Problem Solving)*

Conditioned Response (CR): The response that occurs once a contingency between the CS and US has been learned. *E.g. James has to get regular allergy shots (US) that cause him fear (UR). If the doctor he sees each time wears a distinct lemon perfume (CR), he may form a contingency between the smell of lemons (CS) and his allergy shot (US). If so, the next time he smells lemons (CS), he will begin to feel fear (CR), even if he isn't at the doctor's clinic. (Classical Conditioning)*

Conditioned Stimulus (CS): When paired with the US, creates a learned contingency where the CS (conditional stimulus) reliably predicts some event. *E.g. Every winter, James works at a restaurant with a bell on the door that rings (CS) every time the door opens. When the door opens, the restaurant becomes very cold (US) and James shivers (UR). The next day, James is at a warm spa where he hears a bell ring (CS) and begins to shiver (CR). (Classical Conditioning)*

Confirmation Bias: Our tendency to seek information that supports our hypothesis. One way to overcome the confirmation bias is to actively seek and collect information that may disprove our hypothesis. *E.g. If an individual is opposed to oil drilling in the Arctic, they may naturally tend to collect information about the negative impacts of offshore oil drilling. However, there are also financial, economic, societal and governmental benefits to initiatives like oil drilling that this individual may not consider. (Problem Solving)*

Conjunction Search: A visual search for a target that is distinguished from distracters on the basis of multiple features, such as a combination of shape, size or colour. A conjunction search takes longer than a feature search. *E.g. Finding a green T amongst green L's and red T's. (Attention)*

Conscientiousness: One of the personality traits in the "Big Five" model. This trait reflects an individual who needs to create plans, set goals, and keep their surroundings neat and organized. *(Personality II)*

Consensus: The variable in covariation theory that asks how others would behave in the same way in a given situation. *E.g. If other people would agree with Jim's decision to give presents to his coworkers when he was promoted, the behaviour is likely situational. If other people would not agree with his decision and would not repeat the behaviour themselves, then the behaviour is likely dispositional. (Forming Impressions)*

Conservation: The ability to understand the constant mass or volume of an object using multiple representations. *E.g. A child in the concrete operational would understand that regardless of whether you break a cookie into two or three pieces, you still have the same amount of cookie. (Problem Solving)*

Consistency: The variable in covariation theory that asks whether an individual usually behaves the same way in a given situation. *E.g. If Lisa always comes to work angry and upset, the behaviour is consistent and one could attribute it to dispositional causes. If Lisa usually comes to work smiling and friendly, the angry behaviour is inconsistent and likely to be attributed to situational causes. (Forming Impressions)*

Contextual Cueing: In a visual search, when the expected position of a target is known which helps search more efficiently and find a target more quickly. *E.g. If you are meeting a friend in a lecture in a large lecture hall, they may advise you that they are sitting on the right side of the lecture hall and wearing a blue shirt with a bright pink hat. (Attention)*

Contingency: An association formed between two stimuli after acquisition has occurred. If the presentation of one stimulus reliably leads to the presentation of another stimulus, a contingency can be formed. *E.g. In Pavlov's dog example, a contingency is formed between the sound of the bell and the presentation of food. (Classical Conditioning I)*

Continuous Reinforcement: A schedule of reinforcement in which reinforcement will always follow a particular behaviour. These are infrequent in the real world. *E.g. A dog trained to sit is given a treat every time she obeys (Instrumental Conditioning)*

Control Group: The group in an experiment which does not receive the experimental manipulation to the independent variable and therefore the group to whom the experimental group is compared. *E.g. When testing the effects of caffeine before bedtime on sleep, the control group would not drink any coffee. Researchers would then compare how long it took for these participants to fall asleep, to the group who did drink coffee before bedtime. (Research Methods I)*

Controlled Processes: When one guides his or her attention voluntarily and consciously to an object of interest. *E.g. When studying for a psychology exam, one may actively focus on their psychology notes while music is playing in the background. (Attention)*

Correlation: A measure of strength of the relationship between two variables. Remember, correlation does not equal causation *E.g. A researcher may observe a positive correlation between hat sales and ice cream sales. However, hat sales do not cause more ice cream sales. They both may be caused by a confounding variable, an increase in temperature due to warmer summer months. (Research Methods II)*

Correlation Coefficient (r): Measures the degree to which two variables are correlated. It's value tells us the strength and direction of the correlation. *E.g. Age and height in children have a strong, positive correlation with a value close to r=+1 which means that as age increases, height increases. (Research Methods II)*

Correspondent Inference Theory: An attribution theory proposed by Jones and Davis, stating that we actively analyze a person's behaviour to make inferences or attributions based on three factors. These are **the degree of choice** (whether or not an individual chooses to act a specific way), **expectation** (how likely it is one would behave in a specific way in that situation), and **intended consequences** (what motivation the person may have to behave a certain way because of the resulting consequences). *(Forming Impressions)*

Covariation Theory: An attribution theory proposed by Kelley that is used to determine if a person's behaviour is due to situational or dispositional factors. It is based on three variables: consistency, distinctiveness and consensus. *(Forming Impressions)*

CS-: A signal that reliably predicts the absence of an event (US). *E.g. A green light at a traffic intersection indicates that there will be no traffic from perpendicular lanes, signalling that it is safe to drive ahead. (Classical Conditioning)*

CS+: A signal that reliably predicts an event (US). *E.g. At a railway crossing, the signal of flashing red lights and train bells reliably predicts an oncoming train. (Classical Conditioning)*

Cue: A signal that can be used to trigger the retrieval of a memory. *E.g. Stephen's dad always wears red shirts on Sundays when they watched football together. Many years after he moves away, Stephen sees a man the same height and weight as his father wearing a red shirt on his Sunday, and remembers the times they watched a* *football game on TV. (Memory)*

D

Data: Quantifiable values which measure the outcome of a test. *E.g. When testing the effects of caffeine on alertness, the data would be represented by the number of minutes that each participant required to fall asleep. (Research Methods I)*

Deductive Reasoning: Arriving at a specific fact based on a general theory. *E.g. If you know that dogs are generally friendly, you will be able to deduce that because your neighbour's new pet is a dog, he is likely friendly. (Problem Solving)*

Defense Mechanism: A mechanism by the unconscious ego to keep unacceptable id impulses out of consciousness in order to protect the conscious ego from anxiety. If these id impulses do reach consciousness, defense mechanisms can disguise them, in order to ensure that the conscious ego does not experience anxiety related to these impulses. *(Personality I)*

Deindividuation: A loss of a sense of personal responsibility, individuality and restraint when put in a group situation. *E.g. In the Stanford Prison Experiment, prison guards would perform anti-normative behaviour in a group situation because the anonymity on both sides meant that individual identifying characteristics could not be used against them. (Influence of Others II)*

Denial: A defense mechanism in which the unconscious ego prevents the formation of memories related to the anxiety causing behaviour that the conscious ego engages in. *E.g. A person may deny that they ever were a bully in elementary school, even if reminded that they were, because the formation of anxiety causing memories was prevented. (Personality I)*

Dependent Variable: The variable that researchers observe but do not change during an experiment. The researchers change the independent variable and measure the effects (if any) on the dependent variable. *E.g. When testing the effects of caffeine on alertness, the experimenter observes the amount of time required for each participant to fall asleep (dependent variable) by manipulating coffee consumption (independent variable). (Research Methods I)*

Descriptive Statistics: Describes data in a way that summarizes at a glance the overall results of an experiment. This includes summary statistics such as histograms, measures of central tendency (mean, median, and mode) and measures of variability (standard deviation). *(Research Methods II)*

Developmental Perspective: The perspective in psychology that focuses on how genetic and environmental factors contribute to changes in behaviour across a lifespan. *E.g. A developmental psychologist may examine the effect that parents and peers and genetic factors have on later aggression in teenagers. (Levels of Analysis)*

Discrimination: Restricting the range of CSs that can elicit a CR. *E.g. A certain sound (CS) signals that a phone has received a text message (US) which leads to the individual checking their phone (UR and CR). If another similar sound is repeatedly presented without the arrival of a text message, then the individual would stop expecting a text message in response to the similar sound. (Classical Conditioning)*

Discriminative Stimulus (SD): In instrumental conditioning, it is a cue that signals when a contingency between a particular response and reinforcement is available. *E.g. When a dog is asked to roll over by its owners, it may be rewarded with a treat. However, a vet may not necessarily reward a dog for rolling over. The contingency between the behaviour (rolling over) and the reinforcement (a treat) is only valid in the presence of the owners (discriminative stimulus). (Instrumental Conditioning)*

Displacement: A defense mechanism in which forbidden id impulses are redirected to consciously acceptable targets. *E.g. If an individual has is angry for receiving a bad mark on an exam, it would be inappropriate for them to argue with their professor. Therefore, they may argue with their roommate instead. (Personality I)*

Distinctiveness: The variable in covariation theory that asks whether an individual behaves differently in different situations than the given situation. *E.g. Han always makes time to play with his children, no matter how busy, and this behaviour could be attributed to dispositional or situational causes. If Han makes time to play with and be nice to all children, his behaviour is likely dispositional. If Han frowns and gets upset by other children, his behaviour is likely situational. (Forming Impressions)*

Double-blind Study: An experiment in which neither the experimenter nor the participants know which treatment a particular group of participants is receiving. *E.g. When comparing the effects of drinking caffeinated or decaffeinated coffee before bedtime, neither the participants nor the experimenter would know which type of coffee the participants are drinking. (Research Methods I)*

Drug Tolerance: With each successive drug trial, the drug effects become less and less effective, especially when the drug is taken in the same environment each time. This occurs because the environment reliably triggers compensatory responses, even before the drug has been taken, thus requiring more of the drug to be administered to achieve the same effects *(Classical Conditioning)*

E

Ego: One of the three psychic structures proposed by Freud. The ego is aware of reality and tries to balance or mediate the demands of the id and superego. *E.g. When deciding whether to study or go out partying, the ego would weigh the pros and cons of each option and rationally determine what is the best decision. (Personality I)*

Electra Complex: The female version of Freud's Oedipus complex, proposing that a girl wants to possess her mother for herself, has penis envy and blames the mother for her lack of penis. The girl develops sexual desires for her father, and wants to eliminate her mother, but begins to identify with the mother when she realizes she cannot do so. *(Personality I)*

Encoding Specificity: The finding that memory recall is better when retrieval takes place under similar conditions to encoding. It demonstrates that our memory is affected by our internal and external environment. *E.g. According to this theory, if you were to write an exam (retrieval) in the same room as your lectures (encoding), test performance may be improved because you are retrieving the information in the same environment where you encoded it. (Memory)*

Escape Training: The removal of a negative reinforce following a behaviour in order to increase the frequency of that behaviour. *E.g. If you are consistently getting sunburned while tanning at the beach, you may learn to wear sunscreen to prevent yourself from getting burned further. (Instrumental Conditioning)*

Evolutionary Perspective: The perspective in psychology that explores ultimate causes of behaviour by focussing on how genetic and environmental factors over large periods of time (thousands, if not millions of years across the history of a species) contribute to changes in behaviour. *E.g. An evolutionary psychologist may ask why it makes adaptive sense for children to prefer the voice of their mother to other voices. Does this play an important role in survival? (Levels of Analysis)*

Exemplar Theory: The theory that suggests we categorize new objects by comparing them to all the objects we have encountered in our past experiences. *E.g. When seeing a breed of dog that you have never seen before, you recognize that it is a dog by comparing it to all other dogs you have seen in the past. This involves realizing that this new animal shares similarities with them, such as a distinct nose, a tail that wags and the ability to bark.*

Experimental Group: A group of participants which receives the manipulation of the independent variable, in order to test the effect on the dependent variable. *E.g. When testing the effects of caffeine on alertness, the experimental group would receive regular caffeinated coffee and their alertness would be measured by the time it takes participants to fall asleep. (Research Methods I)*

Experimenter Bias: Actions made by the experimenter, unintentionally or deliberately, to promote the result they hope to achieve. *E.g. When testing the effects of caffeine on alertness, the experimenter may spend more time talking to the experimental group than the control group, thus keeping them engaged and alert in ways not caused by drinking coffee. (Research Methods I)*

Extinction: In classical conditioning, it is the loss of the CR when the CS no longer predicts the presentation of the US. This occurs after repeated trials of presenting the CS without the US. Therefore, the CS no longer leads to the CR. *E.g. If James was repeatedly exposed to the smell of lemons (CS) without receiving his allergy shot (US) and experiencing fear (UR), he would no longer experience fear (CR) when smelling lemons (CS). (Classical Conditioning)*

Extraversion: One of the personality traits in the "Big Five" model. This trait reflects an individual's desire and ease to engage in social interactions, especially in large groups of people. *(Personality II)*

F

False Fame Effect: The finding that subjects who read a list of both newly learned and famous names will incorrectly identify some novel names as famous after a 24-hour delay. This would be due to increased processing fluency, since the novel names would seem familiar to the participant, who would then incorrectly assume that the names seemed familiar because they were famous. *(Memory)*

False Memory: When an individual recalls a fake experience that is remembered as a real experience's from that individual's past. As time passes from initial the description of the experience, subjects become more likely to believe the experience is true. *E.g. A subject may be told to recall the time when they were lost in a park as a child. At first, they may not believe it truly happened, but over time they will be more likely to believe that they really did get lost in a park as a child. (Memory)*

Familiarity: One of the factors influencing attraction. Individuals are more likely to be attracted to things that are more familiar, even if they have only been seen a few times in the past. *E.g. When judging the sound of a baby name, parents may be more likely to like a name that they have previously heard over a brand new, unusual name. However, with more exposure to the name, they may also come to like the unusual name. (Forming Impressions)*

Feature Search: A visual search for a target that is distinguished from the distracters on the basis of one feature, such as colour, shape, texture or size. *E.g. Finding a green T amongst many green L's. (Attention)*

Filter Model: A model that suggests attention works by filtering distractions and allowing only important information through. The subtle difference between a filter model and a spotlight model of attention is that the spotlight model suggests that attention enhances processing of an attended object while filter models suggest that attention helps to ignore other distractions and allow the object of attention to continue for further processing. *E.g. If you wanted to attend to a flower, filter model would suggest that attention helps you ignore the grass around the flower and allow the flower to continue for further processing. (Attention)*

Fixation: In Freudian theory of personality development, when the libido becomes attached to or associated with a specific erogenous zone or object. *E.g. If fixation occurs at the oral stage, than adult sexual may* continue to enjoy oral pleasures, such as eating, more than usual. *(Personality I)*

Fixed Interval: An instrumental conditioning schedule of reinforcement in which reinforcement is delivered after a constant period of time. *E.g. Employees of many companies are paid for every two weeks of work, a fixed interval schedule of reinforcement. (Instrumental Conditioning)*

Fixed Ratio: An instrumental conditioning schedule of reinforcement in which reinforcement is delivered after a constant number of responses. *E.g. In the case of an individual who works on a specific contract, they will be paid after a specific number of tasks have been completed. For instance, an agent for an actress may only receive payment for every 5 auditions he books. (Instrumental Conditioning)*

Fluency: The ease with which an experience is processed. Familiar experiences are generally processed more fluently than are novel experiences. *E.g. You likely process a familiar song with more ease than a brand new song in a foreign language that you have not previously heard. (Memory)*

Flynn Effect: The observation that raw IQ scores have continued to increase every year since 1932. It is argued that this increase may be due to a number of factors, such as increased quality of schooling, accessibility to information, an increased nutrition and health. *(Problem Solving)*

Foot in the Door Effect: When individuals are more willing to comply with a large request after they have already complied with a smaller request. *E.g. If an individual wanted to ask people to make a donation to their charity, they may initially have asked for them to sign up for an email flyer from the same charity. (Influence of Others II)*

Foreign Accent Syndrome: In very rare cases following serious brain injury or stroke, some individuals sound as though they are speaking their native language with a novel, acquired foreign accent. *E.g. A person with a British accent may then sound like they are speaking English with a Russian accent. (Language)*

Formal Operational Stage: The fourth and last of Piaget's stages of development, occurring at 11+ years of age, it is the stage at which a child is able to think in abstract terms. *E.g. A child in the this stage would be able to learn algebra and enjoy fantasy games. (Problem Solving)*

Four Consequences in Instrumental Conditioning: Four mechanisms of describing how reinforcement can be used to influence behaviour. The four consequences are the presentation of positive reinforcement, presentation of negative reinforcement, removal of positive reinforcement, and removal of negative reinforcement. Note: The presentation of negative reinforcement is not the same as the removal of positive reinforcement. *(Instrumental Conditioning)*

Frequency Distribution: A type of graph illustrating the distribution of how frequently a given value or range of values appears in a data set. It is essentially a smooth curve that connects the peak of each bar in a histogram. *E.g. A researcher may create a frequency distribution that illustrates how many students have weights between 120 – 139 pounds, 140 – 159 pounds, 160 – 179 pounds, 180 – 199 pounds and 200+ pounds, respectively. (Research Methods II)*

Functional Fixedness: The difficulty of seeing alternative uses for common objects. *E.g. It is obvious that a flashlight is used to illuminate dark areas. However, it is less obvious or intuitive that you would be able to use it as a paperweight on a very windy day. (Problem Solving)*

Functional Neuroimaging: A non-invasive method in psychology that allows researcher to see what the brain is actually doing while a person performs a particular task. *E.g. A researcher may use functional neuroimaging* *techniques to observe what area of the brain is activated while subjects view a frightening scene. (Levels of Analysis)*

Fundamental Attribution Error: The finding that people often overestimate dispositional causes for others' behaviour, while underestimating situational causes. People are more likely to make this error when determining the causes of others behaviour rather than their own. *E.g. If a cashier is very rude to you, you may think that they are generally a very rude person (dispositional). However, their boss may have just let them know that the company was laying off several people, thus putting them in a bad mood (situational). (Forming Impressions)*

G

Generalization: When a stimulus similar to the CS also produces a CR. *E.g. If a certain sound (CS) signals that a phone has a received a text message (US) which leads to the user checking their phone (UR and CR), then another similar sound may also elicit the same response. (Classical Conditioning)*

Generalization Gradient: A graph demonstrating the extent to which similar stimuli elicit the same CR as the CS in a classical conditioning paradigm. It often appears like a bell shaped curved, in which the original CS elicits the strongest response, and as other stimuli became less similar (further away from the CS), the conditional response becomes less strong. *E.g. If a certain sound (CS) signals that a phone has received a text message (US), this sound will lead to the user checking their phone in response to the sound the majority of times it goes off. Similar sounds may also lead to the user checking their phones frequently and as the sounds become less and less similar to the CS, checking responses will decrease. (Classical Conditioning)*

Genital Stage: The fifth and last of Freud's stages of personality development. This stage takes place at the beginning of puberty when a surge of hormones produces a new wave of libido, marking the beginning of sexuality. The direction of sexual interests depends on where the libido was directed as the stages of childhood progressed. *(Personality I)*

Group Polarization: The finding that group decision making tends to lead to more extreme or polarized views than the original views of individual group members by strengthening the original inclinations of the individual group members. This can move the group decision in either the risky or cautious extreme. *E.g. An individual interested in canoeing is more likely to develop their interest among a group of other individual who enjoy canoeing. (Influence of Others I)*

Groupthink: A group decision-making environment where group cohesiveness becomes so strong that it tends to override a realistic appraisal of alternative opinions. *E.g. A group of city councillors may strongly believe that recycling should be eliminated from their city because it is too expensive and any person who disagrees and tries to discuss the benefits of recycling will be rejected from the group. (Influence of Others I)*

H

Hierarchy of Needs: Proposed by Maslow, a hierarchy of steps of that must be satisfied in order to form a healthy personality. Organized in a pyramid of five steps with associated needs, they are (in increasing order): physiological (what the body needs for survival), safety (a place to live with stability and security), love/belonging (need for close friendships and romantic relationships), esteem (self-esteem and recognition from others), and self-actualization (fulfilling one's potential by maximizing one's abilities). *(Personality II)*

Histogram: Type of graph used to report the number of times a group of values appear in a data set. It forms the basis of a frequency distribution. *E.g. A researcher may create a histogram that illustrates how many students have weights between 120 – 139 pounds, 140 – 159 pounds, 160 – 179 pounds, 180 – 199 pounds and 200+*

pounds, respectively. (Research Methods II)

Homeostasis: The process of keeping the internal state of the body constant, including but not limited to temperature, glucose and ion levels. *E.g. Sweating is a compensatory response that occurs when the body is becoming too warm and needs to cool down to maintain a constant body temperature. (Classical Conditioning)*

Humanistic Approach: An approach to personality that focuses on human interests, values, strengths, and virtues. It is considered to be the most positive approach to studying personality and emphasizes the uniqueness of every individual. *(Personality II)*

Hypothesis: A testable statement that makes specific predictions about the relationship between the variables in a theory. *E.g. When testing the effects of caffeine on alertness, researchers may hypothesize that drinking coffee keeps participants more alert and therefore, it take them longer to fall asleep. (Research Methods I)*

I

Id: One of the three psychic structures proposed by Freud, the id is considered the source of libido. The id is driven by the pleasure principle, causing it to seek pleasure and avoid pain. *E.g. When deciding whether to study or to go out and party, the id would rather go out and party. (Personality I)*

Illusion of the Expert: The perception that something is simple because we are good at it. *E.g. A math professor may have a hard time communicating with his class of first year students because he explains challenging concepts at a level they do understand. He may assume that because he can do the math easily, it must be simple enough for first year students to also understand. (Categories and Concepts)*

Implosive Therapy: A mechanism used to treat phobias, encouraging the individual to confront the CS that causes them anxiety by repeatedly presenting this CS In the absence of the anxiety-causing US. *E.g. Someone who fell from a ladder as a child and now has a phobia of heights may be asked to ride to the top of the CN tower (which results in exposure to heights without falling), thereby treating the phobia (Classical Conditioning)*

Independent Variable: The variable manipulated by the scientist. *E.g. When testing the effects of caffeine on amount of time required to fall asleep (dependent variable), the experimenter can manipulate the amount or type of coffee the participants drink (independent variable). (Research Methods I)*

Inductive Reasoning: Coming to a general theory based on a specific fact. *E.g. If you see your neighbour's dog, Spot, is very friendly, you may use inductive reasoning to determine that all dogs or furry four legged creatures are also friendly. (Problem Solving)*

Infant-Direct Speech: The natural tendency for people to use higher pitch and exaggerated tones when speaking to infants, which may help infants learn to segment speech. *(Language)*

Innate Mechanism Theory: The theory that humans are born with instinctual mechanisms to help them understand universal grammar rules which allows them to develop language skills rapidly. *(Language)*

Inferential Statistics: Statistics that allow us to use results from samples to make inferences about overall, underlying populations. *E.g. If you wanted to use your tutorial to determine the distribution of height of all undergraduate students at McMaster, you would make an inference using the distribution of height in your tutorial. (Research Methods II)*

Intelligence: The cognitive ability of an individual to learn from experience, reason well, remember important information, and cope with the demands of daily living. *(Problem Solving)*

Interval Schedule of Reinforcement: A schedule of reinforcement in which reinforcement is provided once a specific amount of time has passed since the last response was reinforced. *E.g. If a shoemaker is paid for his work every week, then he will receive a pay cheque regardless of how many shoes he has made. (Instrumental Conditioning)*

Inhibition: In classical conditioning, when extinction leads to the learning of a new, inhibitory response to the CS (CS), suggesting the presence of two learned processes (the original learned response to the CS and inhibitory learned response to the CS). *E.g. After extinguishing a fear of bees by repeatedly presenting bees (CS) win the absence of being stung (US), a new learned inhibitory response to bees is learned rather than the contingency between bees and being stung being unlearned. (Classical Conditioning)*

Instrumental Conditioning: The learning of a contingency between a behaviour and a consequence. *E.g. A cat may no longer chew wires if it finds itself getting shocked when doing so. (Instrumental Conditioning)*

L

Language: The human capacity for acquiring and using complex systems of communication. There are three criteria to outline a "true" language. Firstly, language is governed by rules and grammar. Secondly, symbols used in language are completely arbitrary. Lastly, language is productive, meaning there are limitless ways to combine words. *(Language)*

Language Acquisition Device (LAD): An innate mechanism, which Noam Chomsky argues is present in humans of all languages, that helps language develop rapidly according to universal grammar rules. *(Language)*

Language Explosion: The point in children's language development where vocabulary increases very rapidly and most children have mastered the major aspects of language. This typically occurs between one and a half to six years of age. *(Language)*

Latency Stage: The fourth of Freud's stages of personality development, from age six years until puberty. During this stage, the libido becomes channelled into behaviours that are not yet overtly sexual. *(Personality I)*

Law of Effect: Behaviours with positive consequences become stamped in. *E.g. reaching for the cookie jar at the top of the fridge is a positive behaviour if the reward is a cookie.* Behaviours with negative consequences are stamped out. *E.g. if you were blamed and punished for stealing the cookies from the cookie jar, then you probably won't repeat the same behaviour. (Instrumental Conditioning)*

Levels of Processing Model: A memory model, which states that the more deeply a concept is processed, the more easily it is remembered. *E.g. When memorizing "car" on a print advertisement, a shallow level would be remember that the word was spelled in capital letters (CAR). A deeper level would be to understand the word in a context, such as "The Lexus in that ad is driving on a highway similar to the ones my grandfather used to drive home on". The latter level of processing allows the individual to make more connections to the material and remember it better. Overall, the better a concept is organized and understood, the easier it is to memorize and recall. (Memory)*

Low-Ball Technique: An escalation of the terms of an agreement after someone has already agreed. Compliance is secured at a smaller cost, only to later reveal additional costs, making the initial decision seem irreversible. *E.g. An individual may ask their friend to help them carry some groceries. Once they agree to help, they may be asked to help grocery shop, help pay for the groceries, as well as help put them away. (Influence of Others II)*

M

Mean: The average value of the data set, calculated by adding together all of the points in a data set and dividing by the number of items in the set. *E.g. You can calculate your tutorial's mean age by adding up everybody's age and dividing by the number of students in your tutorial. (Research Methods II)*

Measures of Central Tendency: Measures that tell us where a data set is centered, using the mean, median and mode. *E.g. To determine what value age is centered around in your tutorial, you collect all the ages in your tutorial and find that the mean is 18.4, median is 18 and mode is 18. (Research Methods II)*

Measures of Variability: Measures that tell us how other values within a data set fall around the centre of the data. *E.g. Both you and your friend may have the a mean grade in all university courses of B+ but you may have a larger standard deviation than your friend, meaning your grades are more spread out from the mean while your friend tends to consistently get grades around B+. (Research Methods II)*

Median: The centre value in a data set when the set is arranged numerically from smallest to largest. *E.g. If there were 27 students in your tutorial, then the 14th student's age (after age is put in order) would represent the median. If there were 28 students, then the average of the 14 and 15th students' age represent the median. (Research Methods II)*

Message: The facts used by a communicator to persuade listeners, and a contributing factor in persuasion. They can be framed from multiple perspectives to agree with multiple audiences *(Influence of Others II)*

Milgram Experiment: An experiment conducted at Yale University in which a subject was required to shock a confederate learner if they answered a question incorrectly, and progressively increase the shock for each incorrect answer. The experiment measured the willingness of a subject to obey an authority figure when it conflicted with their personal conscience. *(Influence of Others II)*

Mode: The value that appears most frequency in a data set. *E.g. If the majority of students in your tutorial are 18 years old, or if 18 is the most frequently appearing age in your tutorial, then the mode for this data set is 18. (Research Methods II)*

Models: Abstract representations of how the mind functions which can be used to test predictions and design experiments. Models are used in by the Cognitive Perspective and are continually tested and refined to reflect developments in understanding of cognitive processes. *E.g. The Single Memory Model may be used to explain memory processes in the brain. It may then be replaced by a new model when evidence is found to contradict the model. (Levels of Analysis)*

Morpheme: The smallest unit of sound that contains information. A word can contain one or multiple morphemes. *E.g. In the word horseshoes, there are three morphemes. The word is divided as (horse)(shoe)(s), where one morpheme refers to the animal, the next to what goes on its feet, and the last to the fact that there are more than one. (Language)*

Multi-Store Model: The assumption that memory is composed of both short-term and long-term storage systems. In this model, incoming information is first stored in a short-term memory buffer. Information is then transferred into long term memory following active rehearsal of that information. *E.g. When learning lecture material, it is first processed by your short-term memory. Because this only acts as a buffer for that information, it will either be lost or, if rehearsed, transferred into long-term memory for later retrieval. (Memory)*

Multiple Intelligences: A theory, proposed by Gardner, arguing that there are separate and independent intelligences for different purposes. Therefore, one could be brilliant in one area while lacking in another. The eight intelligences that he specified were verbal, mathematical, musical, spatial, kinaesthetic, interpersonal, intrapersonal, and naturalistic. *(Problem Solving)*

N

Normative Function: The role of others in setting norms or standards of conduct based on fear of rejection. *E.g. If an individual doesn't enjoy eating pizza but the rest of the group wants to order pizza, they may conform by expressing the desire to have a few slices to avoid rejection. (Influence of Others I)*

Neuroticism: One of the personality traits in the "Big Five" model. This trait reflects an individual who may experience hypersensitivity and a lot of psychological distress related to fears and depression, therefore requiring emotional support. *(Personality II)*

Normal Distribution: A frequency distribution with a characteristic smooth, bell and symmetrical-curve around a single peak. Many everyday measures are normally distributed. *E.g. Height is normally distributed, with individuals of an average height being the most frequent (creating a central peak) and the shortest and tallest individuals being the least frequent. (Research Methods II)*

O

Object Permanence: The understanding of infants that objects continue to exist even when they are no longer visible. *E.g. In a game of peek-a-boo with their father, an infant has not achieved object permanence if they are continuously surprised every time they see their dad's face again. This is because the baby thinks that their father stops existing when he hides his face behind his hands. (Problem Solving)*

Observational Studies: Studies in the researchers observe the effect of variables they are interested in without performing any manipulation, allowing them to infer correlational relationships but not causation. *E.g. Researchers may compare the number of students who ride the bus and height of snow fallen in meters. This may help researchers determine how to organize the bus schedule based on certain weather conditions without performing any explicit manipulation. (Research Methods II)*

Oedipus Complex: A Freudian conflict that takes place in boys during the phallic stage. Boys want to possess their mother, but see their fathers as competitors that they wish to eliminate. However, the idea of eliminating one's father leads to fear of the father who may retaliate by castrating the boy (castration anxiety). The child resolves this anxiety by identifying with the father. *(Personality I)*

Omission Training: The removal of a positive reinforcement following a behaviour in order to decrease the frequency of that behaviour. *E.g. Billy steals his sister's laptop, which causes their mom to take away his bike for a week. (Instrumental Conditioning)*

Openness: One of the personality traits in the "Big Five" model. This trait reflects a desire for new, exciting and adventurous experiences instead of constantly repeating the same experiences. *(Personality II)*

Oral Stage: The first of Freud's stages of personality development that takes place from birth to one year. The child begins to discover to pleasures of sucking, swallowing, biting and chewing. Some objects first associated with oral pleasure are the mother's breast, bottle or one's own thumb. *(Personality I)*

Others' Opinions: One of the factors influencing attraction. Individuals are more likely to be attracted to those have already expressed that they like them. This is especially true of individuals with low self-esteem. *E.g. You may be more likely to like someone who has expressed their like for you, especially if it was right after you failed your driver's test. (Forming Impressions)*

Outliers: Extreme points distant from others in a data set which easily influence the mean. *E.g. Nine out of ten of your first year grades are between B+ and A+; however, you did not do so well in Chemistry and got one D-. This D- is an outlier in your grades data set and will disproportionately bring down the mean of this data set*

Overdose: When an individual administers too much of a drug, which the body cannot compensate for. It often occurs when an individual takes a drug they have previously taken in a new environment. *E.g. If they have previously taken a drug repeatedly in one specific environment like their room, they may have increased their dosage over time as the drug became less effective due to compensatory responses triggered from their environment. As they moved to a new environment, their body no longer experiences the same compensatory responses, leading to a much stronger effect from the drug. (Classical Conditioning)*

Overextensions: When children apply language rules too broadly, leading to semantic and grammatical errors. *E.g. When a child learns that a TV has a screen that shows images, they may then address a computer, a smartphone, and any other device with a screen as a TV. (Language)*

Overjustification Effects: In order for cognitive dissonance to take place, there must be insufficient justification for the behaviour that is in conflict with the attitude. When overjustification occurs, cognitive dissonance does not happen because the behaviour that is in conflict with the attitude is justified by some external means. *E.g. Emma likes programming code for fun on the weekends. Her friend Jesse though, gets paid to program code and considers it to be very boring. Jesse's behaviour of writing code is overjustified by the fact that he gets paid to do so he does not need to change his attitude to match his behaviour. Meanwhile, Emma's attitude must be that she likes programming code because there is insufficient justification for her behaviour. (Influence of Others II)*

P

P-value: The value resulting from a t-test used to determine statistical significance. If the p-value is below a predetermined threshold (usually 0.05), then the data set is considered statistically significant. *E.g. If you perform a t-test and the resulting p-value is 0.035, you can conclude that your results are statistically significant, meaning there is only a 3.5% chance that you obtained this result by chance. (Research Methods II)*

Partial Reinforcement: A schedule of reinforcement in which a form of reinforcement does not follow every single time a particular behaviour is performed. *E.g. If a young child has a bad habit of sucking his thumb, his mother may reward him with a chocolate for every five days that he does not suck his thumb. Therefore, if the child does not suck his thumb for an entire month, he may receive 4 chocolates instead of 6, as the reinforcement is not provided every time the child performs the desired behaviour. (Instrumental Conditioning)*

Participant bias: When a participant's actions in an experiment influence the results apart from the experimental manipulations. *E.g. When testing the effects of caffeine on alertness, if the participant is aware of the purpose of the study, they may intentionally try to stay awake. (Research Methods I)*

Peripheral Appeal: Well presented, easy to understand messages which are effective for unintelligent audiences. *E.g. When trying to persuade a less academic audience, quick decision making techniques based on heuristics are more successful persuasion tactics. (Influence of Others II)*

Persona: According to Jung, the persona is both an archetype and complex that people have. As an archetype, it is our instinct for social conformity; our instinctual need to be with others and please them. As a complex, it is our public self, feelings, thoughts and impulses that are presented to others because we think they will be approved. *(Personality II)*

Personal Unconscious: According to Jung's theory, it is the repository of an individual's thoughts, memories and emotions that were once conscious but have been repressed into unconsciousness. Complexes are found in the personal unconscious which help make up individual personality. *(Personality II)*

Personality: The combination of qualities, including behavioural and emotional response patterns, making up an individual's distinct character. *(Personality I)*

Personality Factors: A grouping of numerous related characteristics that together compose one specific personality trait. *E.g. An individual may always take the most food at a dinner party, refuse to share, and only talk about him or herself. These characteristics would make up to the trait of being selfish. (Personality II)*

Persuasion: The process of causing someone to believe an idea using reasoning or argument, which may affect their future thoughts, actions and choices. *E.g. An individual may try to persuade their friends to join them on a trip to the Caribbean by talking about how much they all need the break after a hectic semester. (Influence of Others II)*

Phallic Stage: The third and most important of Freud's psychosexual stages of personality development, occurring between 3 and 6 years of age. During this stage, the child discovers the pleasures of stimulating the phallic area. *(Personality I)*

Phobia: An anxiety disorder, in which a person exhibits an exaggerated, intense and persistent fear of certain situations, activities, things or people, which is often disproportional to the danger that is posed. They are often the result of a specific traumatic experience involving the feared stimulus. *E.g. A common phobia is apiphobia (fear of bees). (Classical Conditioning)*

Phoneme: The smallest unit of sound in language, regardless of meaning. *E.g. The word horse can be broken down into three phonemes: /h/, /or/, /se/, each of which has a sound but not necessarily meaning. (Language)*

Physical Attractiveness: One of the factors influencing attractiveness. Individuals are more likely to be attracted to someone they perceive as being physically attractive. This is used on the presumption that what is beautiful is also good. *E.g. People who are physically attractive are often judged as kinder, warmer and more intelligent. (Forming Impressions)*

Piaget's Stages of Development: A set of four basic stages that children actively partake in during cognitive development. The stages are called sensorimotor, preoperational, concrete operational, and formal operational. Each stage is characterized by different abilities and limitations. While different children progress through them at different rates, children must pass through them in the same sequential order. *(Problem Solving)*

Placebo Effect: An effect that occurs when an individual exhibits a response to a treatment that has no related therapeutic effect. *E.g. When testing the effects of caffeine on alertness, a participant may be provided decaffeinated coffee that tastes, smells and looks just like caffeinated coffee. If receiving decaf coffee, theoretically, participants should not have difficulties falling asleep; however, if they do experience difficulty, it may be due to the placebo effect. (Research Methods I)*

Population: The overall group of individuals that a researcher is seeking to study and to which the results of an experiment apply to. *E.g. When testing the effects of caffeine on alertness, a select group of undergraduate students from diverse ethnic backgrounds in Toronto may be tested to represent the entire population of undergraduate coffee drinkers from universities in urban areas. (Research Methods I)*

Pop-Out Effect: A visual search task that is rapid regardless of set size. This occurs when there is a specific feature in the target that highly contrasts the features in the rest of the set. *E.g. A neon yellow square will pop-out in a set size full of red squares. Another example is how a flashing square may pop out in a set containing a large number of multi-coloured squares. (Attention)*

Practice effect: Improved performance over the course of an experiment due to participants becoming more experienced. *E.g. When testing the effects of caffeine on alertness, with repeated trials, participants may* *develop strategies to help them fall asleep, independent of what kind of coffee they are drinking. (Research Methods I)*

Preoperational Stage: The second of Piaget's stages of development, from 2 – 7 years of age, where the child is very egocentric and have difficulty seeing the world from the perspective of others. Children in this stage also begin to master seriation, reversible relationships, and conservation in this stage. *E.g. An egocentric child would think that if they were hungry and wanted to eat, everybody around them must also be hungry and want to eat as well. (Problem Solving)*

Primacy Effect: The finding that memory performance is better for items encoded earlier in a list. *E.g. If you were to repeat a list of words and add a successive word each time, the word at the beginning of the list will be repeated the most. The first few letters of the alphabet "ABC" are more easily remembered than the sequence of letters in the middle "KLO", because the earlier letters are repeatedly more frequently. (Memory)*

Projection: A defence mechanism in which the id's own impulses become attributed to someone else. *E.g. If you are angry at a friend for forgetting your birthday, you may feel guilty about feeling angry and so you project your feelings onto your friend, convincing yourself that your friend is the one who is angry at you. (Personality I)*

Prototype theory: The theory that suggests we categorize new objects by comparing them to *one* internal representation of the category called a "prototype", which is the "average" or "best" member of the category. *E.g. When viewing a new dog for the first time, you may compare it to one typical representation of what you consider a "dog", which is the average of every other dog you have seen before. (Categories and Concepts)*

Proximity: One of the factors influencing attractiveness. Individuals are more likely to be attracted to or become friends with people they work or live closely with. The important thing about proximity is that there is not only high proximity (physical closeness) but also low functional distance (higher probability of interaction). *E.g. You are more likely to be attracted to the friend you sit beside in class whom you talk to and study with than the person who sits directly in front of you to whom you've never spoken. (Forming Impressions)*

Psychodynamic Approach: Freud's approach to personality, which sees personality as generated by internal unconscious psychic structures or processes. These structures determine how we feel and behave. *(Personality I)*

Psychology: The term comes from the Greek word "psyche", which means soul. This field of study emerged as the study of the mind. *(Levels of Analysis)*

Punishment: The presentation of a negative reinforcer following a behaviour in order to decrease the frequency of that behaviour. *E.g. If Sally cheats on a test and is sent to detention (the presentation of a negative reinforcer), this is meant to decrease the frequency of the unwanted behaviour. (Instrumental Conditioning)*

Puzzle Box: A cat trapped in a box with a rope exhibits random behaviours. Its random behaviour would eventually lead it to find that when it pulls the rope, it can escape from the box. Over time, the number of behaviours before the cat finds the rope would decrease, suggesting the animal followed a stimulus-response process, or instrumental conditioning. *(Instrumental Conditioning)*

R

Random Assortment: Assigning participants to either the experimental or control group at random to avoid any biases that may cause differences between the groups of subjects. *E.g. When testing the effects of drinking coffee before bedtime compared to no coffee, the experimenter would randomly assign participants to either condition. (Research Methods I)*

Ratio Schedule of Reinforcement: A schedule of reinforcement in which reinforcement is provided once a predetermined number of responses has been provided. *E.g. If a salesperson sells 4 cars, he receives his pay check. (Instrumental Conditioning)*

Rationalization: A defence mechanism in which the unconscious ego justifies some conscious but dangerous or immoral action. No anxiety is experienced because the conscious ego believes that it has engaged in the said behaviour for harmless reasons. *E.g. If a verbal argument turns into a fight, you may rationalize your physical aggression by believing it was a necessary mechanism of mediating the argument and a form of self defense. (Personality I)*

Reaction Formation: A defence mechanism in which the conscious ego is protected from anxiety by being filled with ideas and feelings that are opposite to the actual unacceptable impulse. *E.g. If a person who has trouble managing their weight experiences an unconscious impulse to order unhealthy fast food, they may find themselves coming up with new exercise plans, diet ideas and promoting healthy eating to all of their friends. (Personality I)*

Recall Test: An memory test in which a subject is asked to read a list of items, and after a time delay, to freely generate as many items as he or she can remember from the list. *E.g. A subject may be required to read a list of 10 animal words and two days later be asked to write down as many words as they can remember from the list. (Memory)*

Recency Effect: The finding that memory performance is better for items encoded later in a list because they remain in the short-term memory stores, since new items do not replace them. *E.g. When trying to memorize a new phone number, you may only be able to remember the last few numbers when you attempt to dial the number. (Memory)*

Recognition Test: An memory test in which a subject is asked to read a list of items and then after some time to read another list of items and determine whether each item is "new" or "old" (i.e. whether it was present in the initial list). *E.g. A subject may be required to read a list of 14 names. After two days, they return and read a new list of 14 names and are required to determine whether these names were present in the list they read two days earlier. (Memory)*

Reductionism: An extreme position within the Biological Perspective in psychology that states that all human behaviour can be explained by reducing a problem solely to the biological mechanisms of the brain. *E.g. A psychologist adopting this perspective may only look at chemical changes in the brain when studying depression. (Levels of Analysis)*

Reliability: A measure of the extent to which repeated testing produces consistent results. *E.g. If an individual performs a certain fitness test multiple times and achieves similar results each time, then the test can be considered reliable. (Problem Solving)*

Representativeness Heuristic: The tendency to assume what we see is representative of a larger category we have in our minds, and to classify people based on how well their behaviour fits with a certain prototype. *E.g. In the gambler's fallacy, individuals believe that previous results influence later games, such as a game of roulette. However, what they do not consider is that each game is completely independent of any past or future games. Therefore, they may assume that a win on one game may represent a larger "winning streak". (Problem Solving)*

Repression: A defensive mechanism in which the ego blocks the id impulse from ever reaching consciousness, but occasionally, some information can slip through. *E.g. A Freudian slip, in which an individual makes a speech error due to some unconscious, hidden idea, is one way that information may not entirely be repressed. (Personality I)*

Research Method: An experimental method used to test a hypothesis. *E.g. When testing the effects of caffeine on alertness, the research method would involve an experiment that compared the time required to fall asleep between a group of individuals drinking coffee and a group of individuals drinking decaf coffee. (Research Methods I)*

Reversible Relationships: The ability to understand relationships from multiple perspectives. *E.g. If a preoperational child is asked if he has a mother, he may say, "Yes, I do! That's my mommy!" If they are asked if mommy has a son, he might say, "Mommy has no son!" (Problem Solving)*

Reward Training: The presentation of a positive reinforcement following a behaviour in order to increase the frequency of that behaviour. *E.g. If a dog is rewarded with a treat for rolling over, he is most likely to repeat that behaviour to obtain the reward. (Instrumental Conditioning I)*

Risky Shift: A theory proposed by Stoner, stating that group decisions tend to be riskier than the mean decisions of the individuals within that group. *E.g. An individual may not want to go cliff jumping on his or her own. However, as a group, everyone may want to. (Influence of Others I)*

S

S-delta: In instrumental conditioning, it is a stimulus that cues when a contingency between a particular response and reinforcement is no longer valid. *E.g. When a dog is asked to roll over by its owners' guests, it will not be rewarded with a treat and therefore, over time will learn not to roll over for guests because the guests (S-delta) indicate the contingency is not valid. (Instrumental Conditioning)*

Sampling: Selecting a subgroup representative of the underlying population to study. *E.g. When testing the effects of caffeine on the alertness of all university students in urban areas, the experimenter cannot perform a test on the entire population and therefore should ensure to select a participant group of university students which is representative of as many ages, genders, and ethnicities as possible. (Research Methods I)*

Scientific Method: A series of steps used to design studies that allow researchers to objectively and consistently find answers to specific research questions. The steps are as follows: theory, hypothesis, selecting a research method, collecting data, analyzing data, reporting findings, and revising theories. *(Research Methods I)*

Scatter-Plot: A graph used to display the relationship between two variables in a data set. Scatter-plots are used to determine the correlation between two variables in a data set. *E.g. If you wanted to illustrate the relationship between age and height, you would plot age on the x-axis and height on the y-axis. (Research Methods II)*

Segmentation: In spoken language, the process of determining when one word ends and the next begins. Segmentation is the process of naturally spacing the sentence into individual words to improve comprehension. *E.g. In this sentence, "Itisdifficulttoreadthisifeverythingisbunchedtogether." (Language)*

Segmentation Problem: The difficulty in segmenting words when hearing an unfamiliar language. *E.g. If you do not speak Polish, it will be very difficult to segment a Polish sentence if listening to someone speaking the language. (Language)*

Self: According to Jung the self is both an archetype and a complex that people have. It is the archetype that drives personality development. As an archetype, it is the instinctive desire for unity, balance, integration and wholeness. As a complex, it is the complex that integrates conflicting and opposing complexes into a unified whole. *(Personality II)*

Self-Serving Bias: The tendency to perceive yourself favourably. *E.g. Even if everybody in a class gets an 80% or higher on a midterm, you are more likely to attribute your success to your own intelligence, instead of simply* *saying the test was easy. (Forming Impressions)*

Semantics: The meaning of each individual word used in a sentence. While a sentence may follow syntactic rules, it can potentially carry no semantic meaning. *E.g. The sentence "Amazing colourless pink flowers jump inside banks" contains no semantic meaning. (Language)*

Sensorimotor Stage: The first of Piaget's stages of development, from birth until 2 years of age. It is the stage at which a child realizes they can affect a change on their environment. Object permanence is achieved at the end of this stage *E.g. A child at this stage may realize that if they push a ball, it will roll across the floor, therefore affecting change on their environment. (Problem Solving)*

Seriation: The ability to logically order a series of objects. *E.g. If a preoperational child is asked to place a series of cones of increasing size order, they may have difficulty ordering them from smallest to largest. (Problem Solving)*

Serial Position Curve: A U shaped curve describing the recall performance of each item in an ordered list of words, indicating that recall is worst for items placed in the middle of the sequence. *E.g. When learning a new unit in a course, information at the beginning and end of the unit will be recalled more easily than information from the middle of the unit. (Memory)*

Set Size Effect: In a visual search, it is the phenomenon of increased difficulty in finding a target amongst distractors as the number of items in the set increase. *E.g. A visual search task in which you are required to find a green T amongst 200 green L's would be more difficult than finding a green T amongst 10 green L's. (Attention)*

Shadow: According to Jung, the shadow is both an archetype and complex that people have. As an archetype, it is the most primitive instinct for sexuality and aggression, but it also acts as a source for energy, vitality, creativity and intuition. As a complex, it is all the things about an individual, including emotions and impulses, that one rejects from consciousness because they are considered "other". *(Personality II)*

Shaping: When a behaviour is too complex for a subject to discover on their own, an experimenter can use shaping by successive approximation. This orders a complex behaviour into a series of smaller, more feasible steps, which when put together, lead to the final complex behaviour. *E.g. An animal trainer may train a dog to carry a toy across the room. First, the dog would be rewarded when they make any movement toward the toy. Once this is reinforced, the dog must pick up the toy in order to be rewarded. Once this behaviour is reinforced, the dog is then rewarded when it crosses the room with the toy. (Instrumental Conditioning)*

Social Facilitation: The increased performance that occurs in the presence of co-actors or an audience. *E.g. A basketball player may score more baskets when playing a real game than when practicing alone. (Influence of Others I)*

Social Learning Theory: Popularized by Bandura, this theory suggests that children learn language imitation and operant condition of adults. *For example, if a baby is babbling and stumbles upon the phonemes to say "mama", they will be praised for it with positive reinforcement, encouraging the child to repeat it again. Parents may also then start repeating this word so that the child will imitate them. (Language)* More broadly, the theory suggests that you learn behaviours by imitating the behaviours of others. *E.g. A child may learn to exhibit aggressive behaviours simply by observing adults acting aggressively. This would not require any explicit reinforcing of the behaviour. (Influence of Others I)*

Social Loafing: The finding that individuals are less motivated when working in a group than when working alone. *E.g. An individual working on a chemistry group project may not work as hard among four other students compared to if they were working alone. (Influence of Others I)*

Socio-cultural Perspective: The perspective in psychology that focuses on how individuals are influenced by culture and interactions with other people. *E.g. A psychologist adopting this perspective might examine how the presence of competitors affects performance in marathon runners. (Levels of Analysis)*

Spearman's G: Spearman's measure of generalized intelligence, which demonstrates that individuals with a high level of "G" would perform well on a variety of classical intelligence tasks and tests. The use of G became controversial when Spearman advocated that only individuals with a minimum level of "G" should be allowed to reproduce. *(Problem Solving)*

Spontaneous Recovery: When a CS again elicits a CR after a rest period takes place following extinction in which the CS is repeatedly presented without the US. *E.g. Sarah is afraid (CR) of bees (CS) after being stung (US) which caused fear (UR). This fear (CR) is gradually extinguished by repeatedly presenting Sarah with bees (CS) without being stung (US). However, the following spring, after long winter of seeing no bees, she suddenly experiences fear (CR) again when she sees a bee (CS). (Classical Conditioning)*

Spotlight Model: A model of attention which suggests that an attentional spotlight enhances processing of objects within the spotlight. *E.g. When searching for your friend in a large lecture hall, you would move your attentional spotlight around the visual scene, enhancing whatever is in your spotlight at that time. You will find your friend when your spotlight reaches her. (Attention)*

Standard Deviation: A measure of the average of each data point from the mean. Data sets with a larger standard deviation are typically more "spread out" than data sets with a smaller standard deviation. *E.g. A population of 30 whose minimum height is 4'8" and tallest height is 6'2" will have a larger standard deviation than one with a minimum height of 5'3" and maximum height of 5'11". (Research Methods II)*

Stanford-Binet Intelligence Test: An intelligence test, developed in the early 1900s, involving 30 short tasks related to everyday life using all types of reasoning. The test is still used in some forms today. *(Problem Solving)*

Stanford Prison Experiment: An experiment conducted at Stanford University to study the behaviour of normal people under situations of authority. Twenty-four male subjects were divided into either prison guards or prisoners. What was supposed to last two weeks was cancelled in six days, because of sadistic behaviour demonstrated by the guards. The experiment demonstrated the issues of authority, assignments of control, as well as the ethical concerns of conducting experiments with humans. *(Influence of Others II)*

Statistical significance: When the difference between two groups is due to some true difference between the two groups and not due to a random variation. *E.g. When comparing mineral content in water found in Hamilton compared to water found in Vancouver, statistically significant data suggests there is actually a difference in the fluoride concentrations, rather than a difference found by random variation or chance. (Research Methods II)*

Stroop Task: A task in which participants are presented with colour words and are asked to name the ink-colour in which the word is presented. Congruent items contain and matching word and colour dimensions, such as RED written in ink-colour red. Incongruent items contain mismatched word and colour dimensions, such as BLUE written in in-colour red. *(Attention)*

Structural Neuroimaging: A non-invasive method in psychology that allows researcher to observe the physical make-up of the brain. *E.g. A researcher may use structural neuroimaging techniques to observe if there are structural differences in the brains of individuals with drug addictions versus those without. (Levels of Analysis)*

Subject/Participant: The individual taking part in an experiment. It may be a human or an animal, depending on the nature of the experiment. *E.g. When testing the effects of caffeine on alertness, a participant will be required* *to drink either decaffeinated or caffeinated coffee and attempt to fall asleep. (Research Methods I)*

Superego: One of the three psychic structures proposed by Freud, the superego is focused on upholding moral principles and forms the basis for one's moral standards and consciousness. *E.g. When deciding whether to study or to go out and party, the superego would rather study. (Personality I)*

Syntax: Refers to the rules that govern how sentences are put together; also known as grammar. Differences in syntactic rules vary according to where the languages originate. *E.g. In French, gender is assigned to all nouns while in English, gender is only assigned to those of biological gender. A table would not be referred to as "it" in English, not "she". (Language)*

Systematic Desensitization: A mechanism used to treat phobias, in which the individual is presented with gradual exposure to the feared CS. An individual would initially be presented with stimuli that are closer to the end of the generalization gradient. *E.g. An individual with a fear of dogs may initially be presented with four-legged animals that do not quite resemble dogs, with each successive step involving conditional stimuli that are more similar to dogs. (Classical Conditioning)*

T

t-Test: A statistical test that determines statistical significance. It considers each data point from both groups being tested to calculate the probability of obtaining differences by chance if there is in fact only one distribution underlying both groups in the experiment. *E.g. A t-test can calculate the probability of finding a difference in height between of members of a basketball team and members of a general population. If the results of a t-test are not statistically significant, this would mean that the members of the basketball team are not statistically different from the underlying population in terms of height. (Research Methods II)*

Taste Aversion: A rapid form of acquisition in classical conditioning. *E.g. Often, it only takes one trial of eating spoiled food and experience of sickness to learn to avoid the food. The CS is taste, the US is sickness, and the UR and CR are aversion. (Classical Conditioning)*

Trait Approach: An approach to personality that focuses on a set of characteristics that define an individual's personality. There are a few proposed models but the most commonly accepted model is the five factor model or "Big Five". *(Personality II)*

Triesman's Dual Filter Model: A model of attention in which there are two filters involved in attention. The early filter is a physical filter that evaluates information based on physical cues such as pitch or intensity. Relevant information is then passed through to the second filter, the semantic filter, which considers the deeper meaning when choosing what stimulus to attend to. *E.g. When listening to multiple musical parts in a song, the early filter may process each musical line. However, the semantic filter will process the meaning taken from the melody. (Attention)*

Tripartite Model: The core of Freud's model of personality, consisting of the id, ego and superego, all representing conscious and unconscious motivating forces in human behaviour. *(Personality I)*

Theory: A general set of ideas about the way the world works. *E.g. A researcher may theorize that certain substances may affect human cognitive abilities (Research Methods I)*

U

Unconditioned Response (UR): A response, which is automatically triggered by a stimulus in the absence of learning. *E.g. If James is exposed to a cold environment (US), he will automatically shiver (UR). (Classical Conditioning)*

Unconditioned Stimulus (US): A stimulus, which automatically triggers a response in the absence of learning. *E.g. If James receives an allergy shot (US), he will automatically experience pain (UR). (Classical Conditioning)*

Underextensions: When children only apply a language rule to a specific context, and do not generalize it to other situations in which they should generalize it. *E.g. A child may use the word car to refer only to his or her family's car, and to no other cars. (Language)*

Universal Phoneme Sensitivity: The ability of infants to discriminate between speech sounds in any language, including those from non-native languages. *E.g. Infants in English-speaking families are able to discriminate Hindi speech sounds almost as well as Hindi-speaking adults while English-speaking adults are unable to accurately discriminate these speech sounds. (Language)*

V

Validity: Measures the extent to which a test is actually measuring what it is supposed to be measuring. *E.g. Conventional eye charts that test vision at different distances may not always produce valid results. A person may have difficulty reading aloud the letters on these charts, despite being able to see the letters perfectly fine, because English may not be their native language or they may have dyslexia. (Problem Solving)*

Variable Interval: An instrumental conditioning schedule of reinforcement in which reinforcement is delivered after some variable period of time around a characteristic mean. *E.g. In the case of an employee's bonus, they will receive it at a random times during some point of the year regardless of how much work they have done. (Instrumental Conditioning)*

Variable Ratio: An instrumental conditioning schedule of reinforcement in which reinforcement is delivered after some variable number of responses around a characteristic mean. *E.g. In the case of slot machines, the payout occurs after a random number of plays around a pre-set mean. Behaviour is reinforced randomly which causes people to play for hours. (Instrumental Conditioning)*

W

Waggle Dance: A form of bee communication that uses physical movement to convey the distance and direction of food away from the hive. *(Language)*

Weschler Scales: Scales standardized to produce an intelligence quotient for every individual. They consist of the Weschler Adult Intelligence Scale (WAIS) and Weschler Intelligence Scale for Children (WISC). An individual who achieves the mean score will be assigned an IQ of 100. Scores around the mean fall along a perfect normal distribution with a standard deviation of 15. *(Problem Solving & Intelligence)*

Whorf-Sapir Hypothesis: The hypothesis that says that language influences our thoughts and the way we perceive and experience the world. *E.g. In a culture whose language does not have a word for war, they do not participate in war or understand the concept of war. (Language)*

Withdrawal: A group of symptoms that occur following the discontinuation of a medication or drug. These symptoms occur most often when an individual is in the environment in which they previously took drugs. These symptoms are internal compensatory response that would normally oppose the drug effect. But without the presence of a drug, these symptoms would bring the body to levels above homeostasis (and lead to discomfort). *(Classical Conditioning)*

Within-Subjects Design: An experimental design in which one group of subjects receives the manipulation of the independent variable to observe the effect of this manipulation on the dependent variable. In this design, one group acts as both the experimental and control groups. *E.g. When testing the effects of caffeine on alertness, there would be two conditions, which involve drinking caffeinated coffee or decaffeinated coffee. One participant would take part in both conditions on separate days. (Research Methods I)*

Z

Zajonc's Resolution: The presence of co-actors or an audience will improve performance in a task in which the individual finds simple or has experience with. It will decrease performance in a task in which the individual is unfamiliar or finds difficult. *E.g. A child may tie his shoes more quickly when his parents are watching but may complete his math homework that he finds challenging when they are watching. (Influence of Others I)*

A Final Word

With the semester drawing to close, it's time to reflect: what have you learned?

The overarching theme throughout our Introduction to Psychology, Neuroscience & Behaviour has been how to apply the scientific method to understand such wide-ranging problems as learning, cognition, social psychology, personality and psychopathology. I hope you have learned to appreciate that the scientific investigation of these problems involves much more than "pop psychology" or even "common sense." In this process, it's important to realize that science does not necessarily seek to prove things to be true. While most people (including scientists) tend to operate as if it did, in fact, what the scientific method does is strongly suggest the things are *not* true. As summarized by Bertold Brecht, "The aim of science is not to open a door to infinite wisdom, but to set a limit on infinite error." Like other sciences, psychology is self-correcting because researchers put forth testable hypotheses that refine our theories of human thought and behaviour.

This means that although many of the core "facts" will be constant in the foreseeable future, some of what you have just learned will undoubtedly change next year! This is the heart of the self-correcting nature of the scientific enterprise. If you have been diligently following throughout this course, the skills you have gained in critical thinking, research methods and evaluating new information will go on with you into the next term and beyond. In PSYCH 1XX3: Foundations of Psychology, Neuroscience & Behaviour, we will build on these skills as we explore processes that guide the behaviours critical to our survival.

As a closing exercise, I would like you to estimate your final grade for this course. Was it above a B or perhaps an A? I doubt that many of you estimated anything in the C or D range. Like the residents of the fictional town of Lake Wobegon where "all the women are strong, all the men are good-looking, and all the children are above average", most of us feel that we are above average on many of the things that we do–above average intelligence, popularity and driving skills, to name a few, not you of course, but perhaps some of your some of your colleagues. However, based on our statistical understanding of average, only part of the population can be above average. Can you think of a good psychological explanation for this effect on our judgment?

Dr. Joe Kim

Degree Programs

McMaster Undergraduate Degree Programs in PNB
Did you love Psychology 1X03? Here are the undergraduate programs that are offered to those students wishing to pursue further study in psychology!

Faculty of Science

Honours B.Sc. Psychology, Neuroscience & Behaviour
The Honours B.Sc. programs are for those students interested in the natural sciences, and thinking about a career in health-related fields, or interested in studying the link between the brain and behaviour.

Honours B.Sc. Biology & Psychology
Honours Biology and Psychology is designed for students with broad interests in the biological and neurosciences. The program is uniquely interdisciplinary. Your studies can merge leading-edge knowledge from both fields, combining molecular biology, development, or ecology with the latest ideas in cognition, learning, or neurophysiology to gain a truly 'big picture' perspective of the living world.

Honours B.Sc. Psychology, Neuroscience & Behaviour (Origins Specialization)
This exciting curriculum is designed to allow students to participate in Origins activities, by interacting with Origins members, taking the Origins courses, and meeting visiting speakers. The specialization is taken as a complement to a variety of participating honours programs, including Psychology, Neurosciences, & Behaviour.

Honours B.Sc. Psychology, Neuroscience & Behaviour (Music Cognition Specialization)
Honours Psychology, Neuroscience & Behaviour with the Music Cognition Specialization is a multidisciplinary program that brings together science and the arts in a unique and innovative way.

Honours B.Sc. Psychology, Neuroscience & Behaviour (Mental Health Specialization)
The Mental Health Specialization is designed to combine teaching and research in psychological, behavioural and neuroscience. This program was developed in response to undergraduate student feedback to prepare undergraduates for graduate and professional training in the area of mental health. This program is unique in Ontario and nationwide. The program will be highly selective and admit approximately 20 students per year from B.Sc. and B.A. majors, with a cumulative GPA of 10+ (A-) or higher.

Faculty of Social Science

Honours B.A. Psychology, Neuroscience & Behaviour
The Honours B.A. programs are for those students interested in fields that focus less on the biological and more on the psychological aspects of human behaviour.

Honours B.A. Psychology, Neuroscience & Behaviour (Mental Health Specialization)
See Honours B.Sc. description for details.

Honours B.A. Psychology, Neuroscience & Behaviour (Music Cognition Specialization)
See Honours B.Sc. description for details

Honours B.A. Psychology and Another Subject (B.A.)
Students may combine Honours (BA) Psychology with another subject from the Faculty of Social Sciences or the Faculty of Humanities provided they meet the entry requirements for both programs. Students graduate with a Combined Honours degree. This is particularly appealing to students who have more than one area of interest.

3-year B.A. in Psychology